"In this timely book, father and daughter Brett and Jessica Finlay distill a wealth of cutting-edge research on aging, health, and the microbiome into an important story on why the trillions of microbes in and on our body (our microbiome) matter for health and longevity. *The Whole-Body Microbiome* is a highly readable perspective on preventing and treating disease, making this a must-read for healthcare providers and everyone who cares about healthy living and aging."
—Alan Bernstein, OC, OOnt, PhD, FRSC, President & CEO of CIFAR

"A fascinating must-read, this well-researched book presents a ground-breaking new perspective to living long and living well.... Who knew, the fountain of youth is full of microbes!"
—Dan Buettner, author of *The Blue Zones*, founder of Blue Zones Project, and National Geographic Fellow

"Revolutionary... The Finlays offer practical, thorough, and sometimes shocking solutions that all of us can implement today, no matter what our age.... By helping our microbes flourish inside our bodies and in our environments, we can find preventive and curative care all in one."
—Dr. Mark Hyman, author of *The UltraMind Solution* and *Food: What the Heck Should I Eat?*

ALSO BY B. BRETT FINLAY, OC, PHD
with Marie-Claire Arrieta, PhD

Let Them Eat Dirt: Saving Our Children from an Oversanitized World

THE WHOLE-BODY

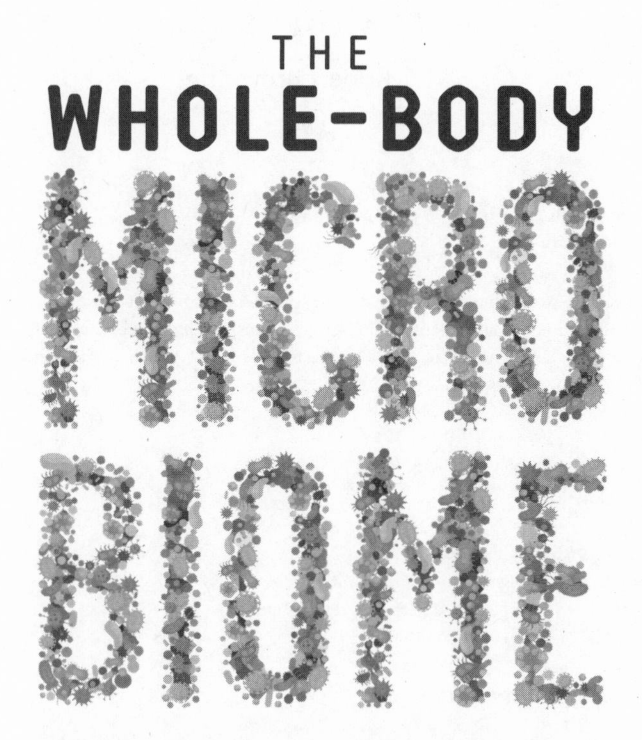

How to Harness Microbes—Inside and Out—for Lifelong Health

B. BRETT FINLAY, OC, PhD
JESSICA M. FINLAY, PhD

Douglas & McIntyre

To Jane and Matt

Grow old along with me!
The best is yet to be…

—Robert Browning

1 2 3 4 5 – 23 22 21 20 19

Douglas and McIntyre (2013) Ltd.
P.O. Box 219, Madeira Park, BC, V0N 2H0
www.douglas-mcintyre.com

Cover and text design by Sophie Appel
Cover illustrations by svtdesigns/shutterstock.com
Author photographs by Jarusha Brown (Jessica M. Finlay) and
Carlos Taylhardat (B. Brett Finlay)
Printed in Canada
Printed on 100% recycled paper

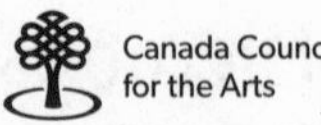

Douglas and McIntyre (2013) Ltd. acknowledges the support of the Canada Council for the Arts, which last year invested $153 million to bring the arts to Canadians throughout the country. We also gratefully acknowledge financial support from the Government of Canada and from the Province of British Columbia through the BC Arts Council and the Book Publishing Tax Credit.

Nous remercions le Conseil des arts du Canada de son soutien. L'an dernier, le Conseil a investi 153 millions de dollars pour mettre de l'art dans la vie des Canadiennes et des Canadiens de tout le pays.

Library and Archives Canada Cataloguing in Publication
Finlay, B. Brett, author
The whole-body microbiome : how to harness microbes–inside and out–for lifelong health / B. Brett Finlay, PhD, Jessica M. Finlay, PhD.
Includes bibliographical references and index.
Issued in print and electronic formats
ISBN 978-1-77162-220-2 (softcover).–ISBN 978-1-77162-221-9 (HTML)
1. Microbiology–Popular works. 2. Microorganisms–Popular works. 3. Bacteria--Popular works. I. Finlay, Jessica M., author II. Title.

QR56.F563 2019 612.001'579 C2018-905939-7
C2018-905940-0

CONTENTS

Preface

"But Dad, what do microbes do *after* the age of twelve–as in, for the rest of our lives? That's what I want to know!" During a family vacation crack-of-dawn run along Hawaii's rugged coastline, we were distracting ourselves from the muggy heat of the rising sun by discussing *Let Them Eat Dirt*. Brett's book, co-authored with Marie-Claire Arrieta, had just come out and presented a somewhat radical idea: that conceiving and raising children in a highly sanitized world is not as good as it sounds. Jessica, Brett's daughter and fellow scientist, was fascinated by the research and findings, but her mind, if not her pace, was already far ahead in the human lifespan, seeking answers for how she could apply microbial science to aging populations, her chosen demographic of study in grad school. Brett was used to this question and knew she was onto something—he had heard some variation of it many times before from colleagues, friends, and public audiences whenever he talked about the book. *I've already had my kids and I am not getting any younger. What about me?*

Brett glanced back after pushing ahead of Jessica up a steep hill. He replied, "Well, we know that gut microbes are involved in asthma, obesity, cardiovascular diseases, and many other serious health issues. But what's really cool is how microbial communities are playing a role in aging in sites distant to the gut across the entire body. Everything from wrinkles to Alzheimer's disease and systemic inflammation has microbial links." Between laboured breaths and the occasional break to take in dramatic ocean views, our wheels began to turn in sync. For the rest of the vacation, when not boogie boarding and hiking (can you tell yet that we're *slightly* high energy?), we bantered about microbes, health, disease, and microbial perspectives on aging. Thus, this book about approaching the end of life was born.

With his sixtieth birthday inching closer each day, Brett already had age on his mind. Peering ahead at retirement, minute reminders of getting older that had previously remained under the surface now permeated his thoughts. Now he brought up the question that he had been pondering since waking up to stiff muscles: "How can I make sure that I'm still running and skiing with my kids–and hopefully grandkids–in my sixties and seventies?" As colleagues and friends slowly but surely began to show signs of aging, and parents advanced through their eighties, discussions about aging had taken a central role in our family's daily life and conversations.

Jessica recalls, "I was thinking about the same thing, but from a very different perspective. When I began graduate school at the age of twenty-two, I began to feel *old* for the first time. Moving to a new city and country with newly minted bachelor's degrees in hand, I felt knowledgeable and much more grown up than the geography undergraduate students I taught. Then a community organization asked if I would consider volunteering with older adults instead of children, and my entire life course shifted. As I made weekly visits to seniors in a wellness outreach program, I took the opportunity to closely observe older people in the community, amassing a data set no longer limited to visits with grandparents. I now noticed the woman with her walker and laden with groceries struggling to cross the street in time, and began waving to the older man who sat on a shaded bench across the street. So often invisible and relegated to the margins, older adults took on a new role at the centre of my life." Now at the age of thirty, Jessica regularly contemplates aging and mortality (even more so than the average thirty-year-old), in addition to those around her. With a doctoral minor in gerontology, Jessica is the resident "aging expert" for her family. Her friends frequently ask for advice on how to preserve their skin and hair as they point worriedly to wrinkles and greys, and happy-hour chatter tends more often than ever before to stray into discussions of ideal places to retire.

The Whole-Body Microbiome is the accumulation of our shared interests in healthy aging and our personal motivation to develop strategies to age *better*. As scientists and researchers, we wanted to see what was "out there" regarding scientific knowledge on aging and the microbiome. After compiling and distilling a massive literature review of existing studies, we extrapolated potential steps we all might take now for health and longevity, and aim to encourage the future direction of research and clinical applications. Along the way, we encountered many questions and misconceptions about microbes and aging, which we highlight in the "Myth/Fact" sections throughout the book.

Despite the title of "Dr." in front of our names, neither of us are physicians, but rather PhDs who are full-time scientific researchers. We are not geriatricians or clinical gerontologists: Jessica's expertise stems from years of studying and working closely with older adults in geographic and public health research, which is a perfect complement to Brett's biological knowledge of the human microbiome. The tips and suggestions throughout this co-authored book are therefore based directly upon academic research; there are no rigid or prescriptive medical recommendations. Expert interviews and key references in each chapter will help readers make informed diet, lifestyle, and personal health decisions.

This book approaches aging as a lifelong process. No one just wakes up one day and finds themselves suddenly *old*. Rather, we accumulate our health, both strengths and risks, over the entire life course. In our shared quest to have more control over aging, we hope this book offers fascinating new ways to harness the microbes everywhere in and around us, so we can enter later life as vital and well as possible.

Age is inevitable. Old is optional.

1

The Fountain of Youth Is Full of . . . Microbes?

From the moment we are born, we begin to die. Aging is a universal but uniquely personal experience. It scares us, bullies us, and motivates us to live better. Because we, as a species, are living longer and longer (more than eighty years nowadays in most developed countries), every one of us has even more time than ever before to grapple with aging and mortality.

Despite what advertisements and doctors may tell us, there is no way to simply "turn back the clock," but we still try to delay the inevitable. We all search for ways to prolong our lives and preserve our bodies—these complex machines made of muscle, bone, and a host of other tissues; our minds, our hearing and eyesight, even our looks. The mythical Fountain of Youth is a spring that allegedly restores the youth of anyone who drinks or bathes in its waters. Since the fifth century BCE, tales of such a fountain have been recounted across the world. Today is no exception—we continue the age-old quest to preserve and restore youth. Instead of seeking the hidden location of the elixir of life, however, we pursue eternal youth through science. Pharmacies, grocery stores, and cosmetic aisles are stocked with anti-aging products, ranging from serums and creams to fight wrinkles and banish spots, to vitamins and supplements promising an elusive "youthful glow."

Scientific studies suggest myriad ways to intervene in the aging process, including antioxidants (to limit the number of free radicals, which cause age-related damage at cellular and tissue levels); calorie restriction (which extends life span and minimizes age-related chronic diseases in a variety of species including rats, mice, fish, flies, worms, and yeast); hormone supplements to treat menopausal symptoms; and a host of dermatological procedures and treatments including retinoids, chemical peels, dermabrasion, ultrasound imaging, laser resurfacing, and cosmetic surgery. While many of these methods have been touted as glamorous and high-tech, one of the most exciting frontiers of current aging science involves the oldest life forms on Earth: microbes.

Contrary to the cutting-edge scientific inventions we're using to make alternative time-reversers, these bacteria have been around for more than 3.5 billion years, from a time when our planet was covered in oceans that regularly reached a boiling point. Our climate has changed dramatically, but microbes are still everywhere: in the air you breathe (they actually made the original oxygen in the atmosphere), on the chair you sit in, and in the food in your fridge. In fact, there are more microbes on your hand than there are people on the entire planet!

Microbes are our constant companions throughout life. Commonly known as germs, they come in many forms, including bacteria, viruses, protozoa, algae, and fungi. While we often blame them for disease (e.g., "I've got a stomach *bug*"), we have only recently realized they are in fact absolutely essential for healthy life. We could not exist without them.

But what do microbes have to do with aging? Everything, actually. We have distinct microbial communities throughout our entire bodies—not just in the gut. These communities affect how our brain, teeth, skin, heart, gut, bones, immune system, and nearly every other body part functions as we progress through life. Well-being is also intimately tied to the microbes that surround us—on our cellphones, kitchen sponges, houseplants, pets, and desks. If you move to a new home or travel abroad, you are exposed to new microbial communities that can disrupt your body's existing microflora (for better or for worse). Your zip code is one of the best predictors of health and longevity, which is underscored by billions of invisible neighbours: microbes. Knowing that there is a continuum between you and the outside world, not a brick wall that ends at your skin, can help you stay healthier over time even if your zip code and age change.

You can harness the microbes in and around you to help keep your gums healthier, bones and muscles stronger, and possibly even protect your brain from Alzheimer's disease and dementia. Of the top ten causes of death in the United States, we now realize microbes can play an integral role in all but one.

MICROBIAL INVOLVEMENT IN LEADING CAUSES OF DEATH IN THE UNITED STATES

1. Heart disease: 633,842 ***
2. Cancer: 595,930 **
3. Chronic lower respiratory diseases: 155,041 **
4. Accidents (unintentional injuries): 146,571
5. Stroke (cerebrovascular diseases): 140,323 **
6. Alzheimer's disease: 110,561 ***
7. Diabetes: 79,535 ***
8. Influenza and pneumonia: 57,062 ***
9. Nephritis, nephrotic syndrome, and nephrosis: 49,959 **
10. Intentional self-harm (suicide): 44,193 *[1]

Data for 2015 obtained from the Centers for Disease Control (2017).
Asterisks (*) denote the authors' evaluation of the strength of microbial involvement.
[1]Microbiota are closely linked to depression, which is a major cause of suicide.

In this book, we will explore and untangle these connections between disease and our bodies' microbes to broaden your understanding of their impact on aging and mortality. Moving from head to toe—or rather skin to brain, mouth to lungs, stomach to gut, etc.—we illustrate how the invisible world of microbes around and inside our bodies is nourishing and essential to a healthy and long life. We think through improvements to lifestyle, diet, and household practices to promote the right kind of microbial exposure. Instead of homes and care environments that are as sterile as possible, for example, we envision places that are comfortable living spaces for us and our microbial roommates. Age-friendly environments need to foster the health and well-being of all inhabitants—microbes included.

Growing Old with Your Microbes

Aging is a natural process that occurs in all biological species, though for some it happens faster than for others. Biologically, we humans hit our prime at around age twelve. In other words, if your physiology (meaning your body and its functionality) remained at that age, you would live more than one thousand years! After twelve, the chance of dying doubles every eight years.

And yet our species somehow beat these odds with increasing success. We spoke to Dr. Anne Martin-Matthews, professor of sociology at the University of British Columbia and former scientific director for the Institute of Aging established by the Canadian Institutes of Health Research. She commented on our current staggering, and unprecedented, aging population: "Over the forty years of my career studying aging, we never anticipated how the field would be shaped by the extended longevity of the population. As recently as a decade or so ago, much of our research focused on 'older people' in their seventies and the 'oldest old' in their eighties. Now we all personally know a 73-year-old with a 95-year-old mother, or an 82-year-old woman concerned about her elderly husband *and* with a 105-year-old parent still alive!"

The advent of vaccinations, antibiotics, and improved sanitation—beginning in the early 1900s—dramatically reduced the number of childhood deaths, as well as deaths due to infectious diseases. This resulted in a major increase in longevity worldwide: Life expectancy increased from thirty-one years in 1900 to seventy-two years in 2016, with Japan's average currently coming in highest at eighty-four! Dr. Martin-Matthews noted that the number of centenarians (older than 100) continues to increase worldwide, along with more supercentenarians (older than 110). This is one of the most significant social transformations of the twenty-first century.

While virtually every country in the world has growing numbers of older people, chronic ailments such as obesity, type 2 diabetes, asthma, and inflammatory bowel diseases are also rapidly on the rise worldwide. Far from being limited to developed regions of the world, chronic diseases are accelerating in developing countries. The number of people in the developing world with diabetes, for example, will increase by more than 2.5 times, from 84 million in 1995 to more than 228 million in 2025. The World Health Organization (WHO) estimates that the global burden of chronic disease will rise to 60 percent by the year 2020, from approximately 46 percent in 2001. Almost three-quarters of all deaths in 2020 will be attributable to chronic diseases.

These conditions plague many individuals' health and reduce quality of later life. In terms of incidence—i.e., the number of people affected—cardiovascular disease and brain diseases such as Alzheimer's, Parkinson's, and dementia are on a dramatic upward trajectory in our society. But despite more than twenty years of intense research, Dr. Martin-Matthews reflected there is still no "silver bullet" to address Alzheimer's disease and related dementias: "Research has primarily focused on the brain and its structural changes. Yet the reality is that we have not been able to find effective prevention or treatment measures. Multiple billions of dollars of research investment later, we're coming to understand that the issue is much more complex. We clearly need to look elsewhere. Perhaps the microbiome plays a role—maybe to solve problems in the brain we need to look at the gut and other areas of the body where microbes are involved." Dr. Martin-Matthews admits she is no microbe expert—quite the opposite, actually, as she confessed at the beginning of our interview.

And yet, as a sociologist and social scientist she is very clued-in to possible microbial interactions. Microbes affecting our brains, as well as other distinct bodily and environmental sites, are ripe for investigation given that they offer immense potential to better understand aging. Researchers with the openness of Dr. Martin-Matthews have already drawn remarkable connections between the microbiome and conditions such as obesity, type 2 diabetes, asthma, and inflammatory bowel diseases, in one way or another. They have also found links to many other normal physiological changes associated with aging, such as loss of bone and muscle mass and skin wrinkling. This research involves microbial effects well beyond the stronghold of our guts in important body sites such as the brain, heart and bones, as well as critical human environments like hospitals and nursing homes. By better understanding our everyday environmental microbes, we believe we may be able to strategically manipulate them so that we can live healthier and longer lives.

In Sickness and in Health: Microbes, Our Lifelong Partners

We are inhabited by more microbes in and on our body than we can imagine: There are at least as many bacterial cells in the body as human cells. They surround us and cover us. There are actually layers of microbes between every body surface and the environment, not only between your skin and the pages of this book, but also between the air you breathe and your lung cells. Normally we all coexist peacefully, but our microbial interface is heavily influenced by changes in our environment. Every moment has the potential to radically alter the landscape of our microbiome—and thus impact health.

Microbes are especially important at the bookends of life: the first few years of childhood and the last years of adulthood. Cohabitation with microbes begins with our journey through the birth canal, when we consume mouthfuls of bacteria. We then get a regular dose of microbes from our mothers by drinking breast milk and being held skin-to-skin. This critical microbial bolus jumpstarts our immune system and impacts brain development as we set off on our lifelong relationship with microbes. As a result, by the time we reach adulthood, there are over five hundred distinct species living at any one moment in our intestines. The microbiome is fully formed by the age of two to three.

Microbes help break down food in the digestive tract and harness energy and nutrients. They keep our immune system functioning and help protect against pathogens that we constantly come into contact with. As we age, however, the role of microbes changes. People over the age of seventy have radically altered microbial communities from when they were younger. The composition of microbiota also shifts, which may lead to detrimental effects on older people. For example, as we will see later, age causes an increase in inflammatory microbes and a decrease in helpful microbes that dampen the immune system. Collectively, this results in an increase in low-grade inflammation throughout the body causing tissue damage, a process called inflammaging. These changes can lead to greater susceptibility to diseases and a general decline in health. Knowing that microbes are central to the process at the heart of general body decline is a great discovery for science—and for all of us—and highlights the critical need to maintain and enhance our microbes as we age.

It is estimated that longevity is 25 percent genetic and 75 percent environmental. This is a profound statement: It means that we have the ability to control the majority of elements that affect our health and lifespans. Just because your parent had cancer doesn't mean that your fate is sealed—although you may have genes that increase your likelihood of getting it. Your future is affected much more by your environmental exposures. We now realize that when we talk about the "environment" as being that which is in closest contact with our body, we actually include microbes.

It is natural for our microbiome to decline as we age, but this process has undergone a radical shift because we've turned our environment into a battleground for our modern crusade against "germs." Our common understanding that some bacteria and viruses make us sick, and should be avoided, is very true. These are the notorious germs we continually battle—pneumonia and flus and skin infections. But we've taken that truth to an extreme that is causing us to suffer dire consequences, speeding up and exacerbating the annihilation of key microbes we likely need for longevity.

As far back as the nineteenth century, pioneering microbiologists focused on chasing the microorganisms that caused infectious diseases, including cholera, tuberculosis, and diarrhea. Scientists discovered that "germs" caused rabies, anthrax, and other infectious diseases, so the hunt for disease-causing microbes escalated. Then everything changed around the middle of the twentieth century. We developed antibiotics to rid ourselves of some disease-causing infections. While these wonder drugs have saved millions of lives, they have also led to our modus operandi to kill microbes—all of them—rather than try to learn more about them and live in harmony, as we had for most of history. Other inventions, such as hand sanitizer, antimicrobial mouth washes, and household antibacterial cleaning solutions became everyday products. "If clean is good, then cleaner must be better!" we shouted from every rooftop, billboard, TV commercial, and transit station. High on our own scientific mastery over these tiny bugs, we didn't realize that we were sending ourselves toward current disease epidemics.

A glance around our society today reveals a completely different set of diseases than those of a hundred years ago, when infections were a major killer. Obesity, diabetes, inflammatory bowel disease, allergies, and asthma are but a few examples of conditions we now realize are partly collateral damage from our war on microbes. This is because antibiotics do not just kill off the "bad" microbes, they wipe out the "good" ones, too. This is why using antibiotics can cause diarrhea, an upset stomach, and even urinary tract infections. Taking these powerful drugs will kill off the agent of disease, but they will also leave our body vulnerable in other ways by eliminating the good microbes that protect us. We call it the "Hygiene Hangover," or the head-splitting price we pay for our century-long bender of antimicrobial precautions.

Furthermore, our highly sanitized world remarkably reduces the overall number of microbes that we are exposed to. Every generation has fewer types of microbes in and on their bodies than the previous one. More and more of us spend our days inside sanitized and climate-controlled buildings. Our bodies' ability to work as they were designed to, in the presence of abundant and diverse microbes, is being compromised as a result, jeopardizing a key part of continued human evolution, let alone personal health.

We evolved over millions of years with these invisible partners, and yet within two to three generations, many of them could be on their way to extinction.

The Book of Microbes: Charting Uncharted Territory

What to expect when you're expecting to grow old? While there is an entire industry devoted to preparing expecting parents, scant resources guide us on how to grow old. Every day, more and more young adults, professionals, Baby Boomers, and family members with aging loved ones actively seek out advice and information. We mainly turn to online resources or other sources of often-outdated science on aging. The supply of legitimate and scientifically based resources indeed remains scarce—and is a key reason why we're writing this book. What we present here is not an exhaustively full or definitively complete picture of the specific role of microbes in aging; instead, our goal is to provide a starting point for everyone who appreciates the fact that well-being is a lifelong process and wants to improve their health as they age.

In each chapter we explore the fascinating and invisible world of microbes and its effects in each major system or organ of the body, and present the current scientific understanding of how the systems' microbes might affect aging, from head to toe. We also offer lifestyle strategies and "Quick Tips" that we can all take advantage of, whether we're eighteen or eighty, to grow old as healthfully and gracefully as possible. Leading medical professionals and scientific researchers lend their voices throughout to deepen our grasp of how knowledge of the microbiome is influencing their respective fields. Most are not microbiology experts—rather, they represent medical doctors, dentists, biomedical researchers, public health experts, and social scientists all grappling with their own well-being in addition to their work.

Both of us are scientists, and all the material presented in this book is based on the current scientific literature. Each chapter includes a list of select scientific papers that helped shape our knowledge. These are not exhaustive lists—we read over one hundred papers for each chapter—but the papers and reviews highlight major concepts in the chapter. If you are interested in looking further at the scientific literature, we encourage you to use the PubMed website (ncbi.nlm.nih.gov/pubmed), which offers search options for biomedical literature and provides abstracts of peer-reviewed articles and reviews. Some articles are open to the public, while other journals charge for access to the full article, although libraries can usually get the full articles to you for free.

We will cover several key concepts in this book. First, we will share exciting new data coming out in microbiology that will rewrite many chapters of modern medicine. However, the field is still very new—barely a decade old—which means that there are more questions than answers, especially when it comes to direct applications. And as with any new scientific field, the data can be full of conflicting findings and rather confusing results, making it harder to parse. This is especially true of microbiology because of the differences found between individuals. However, there are also common themes that are emerging from all these studies that we highlight, such as beneficial health effects from having diverse gut flora, and how the microbiome influences inflammation at various body sites. Our goal is to synthesize this overwhelmingly large body of information and condense it into a digestible (pun intended) format that will be of use in your daily life.

Second, we focus our attention on the microbiome's effects on healthy aging—a perspective that has not of yet been covered in popular literature. We believe this is a critical gap as there are many concepts in this book that can help you age with relatively good health by understanding our microbiome. At first, you might think this means only individuals who are over, say, the age of sixty-five. However, healthy aging spans one's entire life. What you do during your twenties can have effects decades later. Remember that ankle you sprained at twenty-two playing basketball, which is now giving you a hard time at seventy? This book is written for people of all ages who care about their long-term health.

Third, as we discussed above, there are microbes all over, inside, and around our bodies. Nearly all the attention so far has been on gut microbes. There is no doubt that they play a major role in all sorts of body functions. However, we are now beginning to realize that the microbes in different sites of the body also contribute in individually fascinating ways and play a large role in health, disease, and aging. Large microbial communities live happily in the mouth and oral cavity, the skin, and the urogenital tract—and, like the gut microbes, they busily make chemicals and do other things that can affect our body's functions. A recent study, for example, found that the microbiome contributed to over a third of the differences in people's HDL/cholesterol levels, and a quarter of the variation in body mass index (BMI). We can apply this knowledge toward developing effective personalized diets based on a person's unique microbe makeup.

Throughout the book we cover many areas of the body to better understand aging processes and consider specific lifestyle and nutrition interventions to enhance healthy microbiomes. There is some slight overlap between chapters when there are clear interactions among body sites, such as the gut microbiome's effects on the brain. We even discuss the microbes surrounding our bodies, such as those in your home or on your cell phone, which also immensely impact our own microbiomes. North Americans, on average, spend approximately 90 percent of their time indoors. Current cleaning and hygiene standards in private and institutional settings—especially hospitals and nursing homes—tend to promote multi-resistant pathogens instead of supporting beneficial microbes. In the absence of contact with these key microbes, some dynamic microbial lineages have decreased in prevalence and abundance in today's humans. For this reason, the intestinal microbe chapter doesn't come until midway through the book. We want to stress the importance of the whole-body microbiome to healthy aging, not just the gut microbes we hear about nearly every day.

By reading this book, you are about to embark on an exciting journey into a brand-new universe of human health and disease. The microbiome has been called the "most recently discovered organ" and its many functions are only now being uncovered. It is our hope—and the hope of all the scientists working in this area—that we can use microbiome science to alter that top ten mortality list, or at least decrease the numbers of people with those diseases.

We guarantee you won't see the world the same way after you finish this book. You may pause and consider before shopping for groceries, eating lunch, brushing your teeth, sending a text message, washing your hands, or scrubbing the dirt off a freshly-plucked carrot. We hope the journey is as fun and enlightening for you as it has been for us, and that we will all benefit from embracing microbes as long-life partners.

2

Your Microbes Are Glowing: The Skin Microbiome

We all want to live long, healthy, and fulfilling lives. But we don't exactly want the *long* part to show on our face and skin. From hair dyes and anti-wrinkle creams to Botox and surgical facelifts, it seems that there is no limit to the ways we try to preserve our appearance. With age, the skin becomes more fragile and more vulnerable to damage. Despite the staggering number of commercial products and DIY skincare concoctions available, anti-aging therapies often lead to mixed results. Attempts to "turn back the clock" on sun damage and minimize wrinkles and age spots in general have plateaued. Retinoids—considered the best anti-aging products available—have been around since the 1980s, and the pace of innovation for other topical products such as moisturizers and serums has slowed. A new approach to aging skin and its care is sorely needed.

For years, conventional wisdom dictated that we use antibacterial agents to blast "bad" bacteria, which was partially the cause of blemishes, aging, and illness of the skin. However, researchers are discovering that these toxic antibacterial products are, in many cases, far more dangerous to overall skin health than any bacteria they are designed to kill. We now realize that microbes have a critical, and most often positive, role in skin health: They can reduce wrinkles, dryness, sun damage, resistance to infections, acne, and even body odour. Compared to the killing effect of antibacterials, microbes keep our skin healthy and strong: They alter and educate our immune system, and out-compete certain disease-causing pathogens. The technical term for this is "competitive exclusion," meaning that the resident microbes occupy any potential colonization site available to pathogens and try to out-compete those pathogens for nutrients and places to bind. As a result, there is a lot of potential for microbial strategies to rejuvenate our skin in adulthood, even after the damage is done. A better understanding of how these microbes work offers new ways to age gracefully inside *and* out.

Beauty Is Skin Deep

We are completely coated in microbes. There are approximately one million bacteria composed of hundreds of species on every square centimetre of our bodies. Skin contains the body's fourth-largest collection of microbes after the gut, oral, and vaginal areas. Have a look at your hand—there are over 150 species on one palm alone. Although we tend to think of our skin as homogeneous, from a microbial point of view it comprises immensely different environments: Both the moist, warm areas—the "tropics" of your groin, armpits, navel, and between the toes—and drier surfaces—the "deserts" of your forearm, buttocks, and hands. Although we don't know why, the dry areas of the skin have the most diverse composition of microbes but lower microbial numbers. (Just like plants, microbes need moisture to grow.) There are even microbial differences between the sides of our bodies. One study found that we share 68 percent of microbes between the left and right forearms, but only 17 percent between the left and right hands. While alarming at first, this difference actually makes perfect sense: In the last five minutes, which hand did you use to pick up a pen, scratch your head, or tap out a message on your phone? The things our two hands touch vary wildly, whereas our forearms are used in a much more even capacity.

What do these teeming colonies of surface microbes do to us? In addition to out-competing harmful microbes, they also produce molecules that counter specific pathogens. For the most part, they go about this business silently, breaking down fats and other molecules produced by our skin for food so that they can be our strongest first line of defence. To do this well, they also interact with the immune system, reporting on the state of our skin surface, so the immune system can better respond to and defend against foreign invaders.

One microbe that illustrates these concepts is *Cutibacterium acnes* (*C. acnes*, previously known as *Propionibacterium acnes*), so named for its involvement in acne. This microbe breaks down the triglycerides (fats) in sebum, an oily secretion that is produced by our skin glands. Breaking down these molecules produces fatty acids, which in turn acidify the skin. This acidity is necessary to block pathogens such as *Staphylococcus aureus (S. aureus)* from causing infections on the skin, as the bacterium prefers a more neutral pH. In fact, people with atopic dermatitis (an inflamed skin disease) have lower levels of beneficial microbes and higher levels of pathogenic *S. aureus*.

Another common skin inhabitant, *Staphylococcus epidermidis (S. epidermidis)*, works for our benefit by secreting molecules that kill pathogens like *S. aureus* and Group A *Streptococcus*—pathogenic bacteria that can cause a variety of illnesses, ranging from minor skin infections and medical conditions such as pimples, cellulitis, atopic dermatitis, abscesses, and strep throat to life-threatening diseases such as pneumonia, endocarditis, and sepsis. A recent study showed that certain normal skin microbes, including *S. epidermidis*, secrete these antimicrobial peptides in order to kill off the competition vying to live on the skin. The researchers found that when these beneficial microbes were added to the skin of patients for twenty-four hours, the number of *S. aureus* (the pathogen) decreased. The takeaway from all this seems to be that we can actually benefit from this ongoing microbial warfare on our skin. But if we lose the beneficial microbes, we also lose their protection.

While you wouldn't want to go around guessing people's ages by their faces, from a microbial perspective it's completely possible to do so. Because microbes are our essential companions as we age, and respond to our ever-changing internal and external environments, we can tell someone's age within a decade just from analyzing a microbial swab of the forehead. Remarkably, people over the age of fifty have distinctly different microbial signatures than younger adults. Scientists are just beginning to uncover exactly how and why the skin's microbiome shifts and loses diversity as we age. No matter the reason, this phenomenon speaks to the critical need to enhance and maintain your skin microbiota as a robust ecosystem throughout life.

Cosmetic companies have picked up on this fact: You can now find commercial skincare products that incorporate emerging microbial scientific discoveries into topical applications. At the time of writing, L'Oréal, for example, patented several bacterial treatments for dry and sensitive skin; Estée Lauder patented a skin application with *Lactobacillus plantarum*; and Clinique sells a foundation with *Lactobacillus* ferment. Products such as La Roche Lipikar Baume AP, used to treat eczema and other dry skin issues, likewise include bacterial additives to help restore a healthy skin microbiome and to stop itching. Keep an eye out for these and more emerging microbial skincare lines.

Despite this progress, Dr. Greg Hillebrand, a senior skin scientist at Amway, a major health and beauty corporation, believes there is still a serious need for new methods and treatments for aging skin. "The pace of innovation in the anti-aging category is slowing. Conventional topical products like moisturizers, serums, and essences contain active ingredients aimed at preventing or reversing the signs of aging. Retinoids [a class of active ingredients] remain the gold standard, yet they have been around since the 1980s. The skin microbiota represents an exciting new focus area for us, and it's the next best opportunity to solve many of the challenges associated with aging skin." Dr. Hillebrand's enthusiasm for the use of microbes goes back to 1995. He was sent to Japan by his former employer, Procter and Gamble, to figure out exactly how it worked by studying a prestige skincare line that consisted of a concentrated fungi ferment cultivated, processed, and filtered down into an essence product. "Many of my colleagues at the time did not actually believe it did anything; they all thought it was 'foo-foo dust.' I was only there for a few months when my director from the US came over to see how I was doing. I was excited to share my progress and ideas and met with him and my VP. I told them that I thought it might be possible that the fermented filtrate worked in part by favourably modulating the bacteria on the face in a way that we didn't yet understand. Basically, I was proposing that the use of the product might shift the bacterial composition, perhaps maintaining the good ones and not the bad ones on the face." In 1995, we didn't yet appreciate the concept of "good" and "bad" bacteria on the skin—it simply wasn't conceivable that the skin microflora were important; we certainly didn't culture bacteria *specifically* to benefit the skin—so it was not surprising that the reaction of Dr. Hillebrand's director to this novel idea was less than enthusiastic.

Thankfully, Dr. Hillebrand followed his hunch. Flash forward to present day, when he is critically involved in clinical studies and product testing for Amway explicitly focused on the microbes. Much of his team's effort involves data gathering to better understand this emerging area. Amway set up a clinical test site, for example, during a major art event in Grand Rapids, Michigan, where they figured there would be a large variety of patrons of all ages in attendance. The Amway team measured the skin microbiome of hundreds of festival-goers via swabs of their scalps, foreheads, forearms, and nasal/oral areas. The samples showed fascinating differences in microbial communities, especially associated with people's ages. Describing this particular sampling, Dr. Hillebrand grew animated at the prospect of how we can use these microbial differences to enhance the appearance and health of aging skin.

While gaining traction, he admits that the concept of embracing, rather than eradicating, bacteria on skin remains as foreign today as the discovery of groundbreaking topical skin products was thirty years ago.

But he remains optimistic: "With the microbiome, we can actually *do* something. The challenge now is to figure out exactly how to leverage science into more effective skin products for real innovation."

Skip the Antibacterial Soap

One of the few certainties we have when it comes to the microbiome—and a common theme throughout this book—is that we need to become *less clean* in our everyday practices. A fine line exists between being hygienic and over-sanitized. When caring for sick or elderly people, hand sanitization is a common, and often mandated, practice. Hand sanitizer dispensers dot the halls and carts of nursing homes, hospitals, and clinics, where spreading bacteria could wreak havoc on patients who have a variety of serious illnesses and compromised immune systems. Every day in the United States, about one in twenty-five hospital patients has at least one healthcare-associated infection—*while* they're being treated for something else.

But we're often, and unnecessarily, encouraged to use antimicrobial cleaning supplies and disinfectants in daily life to prevent the spread of infection. Visit any high-traffic tourist area or restaurant and you'll be blasted with squirts of astringent hand sanitizer. According to a recent US Food and Drug Administration (FDA) statement on the safety and effectiveness of antibacterial soaps, hand sanitizers do not uniformly prevent the spread of germs, as we all thought. In fact, the FDA ruled that companies can no longer market "antibacterial" washes containing certain ingredients (the most common being triclosan and triclocarban) because manufacturers did not sufficiently demonstrate that the ingredients are effective for

MYTH: Antibacterial soaps are necessary to sanitize your hands.

FACT: You do not need to use antimicrobial soap to get the job done: Using regular household soap with proper handwashing technique is generally a better way to remove germs.

long-term daily use. Repeated antibacterial hand-washing seriously disrupts the skin microbiota, which can result in various skin irritations, disorders, and infections. Some data even suggests that these antibacterial ingredients may do more harm than good over the long-term—for individuals and the whole population—by creating hardier, more resistant bacterial strains.

Alcohol-based hand sanitizers kill most of the microbes on your hands, good and bad. However, they cannot kill spores of *Clostridium difficile (C. difficile)*, a common pathogen in healthcare settings that causes severe diarrhea and other potential complications. To ward off these bacteria, regular soap and water is all you need! With a *C. difficile* infection, a healthcare provider should wear gloves to examine patients and wash his or her hands before and afterward; wearing gloves alone is not enough to prevent the spread of infection. Surprisingly, according to the US Centers for Disease Control and Prevention (CDC), healthcare providers clean their hands less than half of the times they should. Patients and loved ones can play a role by both politely reminding healthcare providers to clean their hands and serving as a good example by doing so themselves. This simple step is generally the best way to avoid getting sick and spreading germs to others.

THE CDC RECOMMENDS HAND-WASHING

- **Before preparing or eating food**
- **Before touching your eyes, nose, or mouth**
- **Before and after changing wound dressings or bandages**
- **After using the bathroom**
- **After blowing your nose, coughing, or sneezing**
- **After touching hospital surfaces (e.g., bed rails, bedside tables, doorknobs, remote controls)**

If soap and water is not an option or you have to use a hand sanitizer—such as in an elder care home where this practice is mandated for employees–the CDC recommends that it be an alcohol-based hand sanitizer containing at least 60 percent alcohol. Put the product on your hands and rub them together, covering all surfaces until the hands feel dry. This should take approximately twenty seconds. The FDA's antibacterial soaps ruling does not apply to healthcare settings, so be sure to inquire about the use of these products when in these locations.

An Eye for an Eye

Why are contact lens wearers at risk for frequent eye infections? It may have to do with the delicate balance of our ocular microbiome and what happens when it comes into contact with the skin. If we displace the eye's microbiota through bacterial organisms introduced on contact lenses, this could affect the eye's ability to fight off infection. Brett conducted such an experiment in grad school: He washed one of his contact lenses with the usual cleaning solutions and enzymatic digestion—how they were cleaned many years ago—and put the contact lens under a powerful scanning electron microscope. It was a terrifying sight! Instead of having a nice pristine lens surface, the contact was coated in a thick blanket of microbes. To top it off, this was the first biofilm Brett had ever seen; needless to say, it's an image he will never forget—especially when putting his contacts in every morning.

There is also a practical link between contact-related infection and age. Because eyesight shifts with age, and many of us rely on contact lenses instead of glasses to correct vision, our microbe-coated hands and eyes have more chances to swap bacteria. In an experiment of twenty volunteers, researchers found the ocular microbiota of contact-wearers more closely resembled that of the body's skin rather than the standard eye microbiota. In other words, constant blinking and tears—which contain natural antimicrobial substances—keep our ocular microbiota load low most of the time. Regularly touching our eyes can shift this balance.

The take-home message is to be aware of the microbes living in the moist biofilm of contacts. Under normal conditions the eye's surface can overcome these pathogens; however, trauma, surgery, disease, and, notably, age make eyes more vulnerable. Washing one's hands before applying or removing contacts helps immensely, as does cleaning the contacts and storage case thoroughly with contact lens solution (which contains substances that inhibit microbial growth). Don't think you're off the hook if you are only an occasional contact wearer: Organisms transferred to the lens from the skin during removal that are not destroyed by the disinfecting solution can proliferate over time in a static lens case.

Scientists have successfully applied antimicrobial coatings to biomedical devices such as artificial hips and other objects that are implanted in the body, so a logical next step is to apply this to contact lenses. Currently there are several options commercially available, and animal models offer promising evidence that antimicrobially treated contact lenses can decrease risk of eye infection. It's unclear, though, how these coatings will affect the eye microbiota as a whole.

Simply Sebum

Rub your finger along the crease of your nose: feels oily and greasy, right? Despite popular misconception, this layer of oil is normal. It's a substance called sebum, which the body secretes out of hair follicles to keep the skin moist and supple. If sebum becomes trapped in a hair follicle it can lead to a buildup of the acne-causing *C. acnes*. Often the bane of teenage years, acne can unfortunately flare up again around menopause for women, due to changing hormone levels. Whether fifteen or fifty years old, the root cause is the same: changing hormone levels. As relative testosterone levels rise, the skin's sebaceous glands can go into overdrive and produce excess sebum. But whereas things eventually level out once puberty ends, in older women the problem is exacerbated by slower cell regeneration and prolonged buildup of *C. acnes*. Frustrating acne blemishes can pop up near the chin, jawline, and, sometimes, upper neck. Unlike the superficial zits teens get on their T-zones, these blemishes are often more like cysts, smaller and more tender deep below the skin—hence their being more painful and difficult to remove.

Thankfully, this pimply condition diminishes once a woman settles into postmenopausal hormone levels. In the meantime, skin should be treated kindly: do not strip it with a strong astringent if the skin is not oily. Probiotic applications that supplement the skin's microbiota with a beneficial microorganism might offer promising treatments for older women's acne. One study found that *S. epidermidis* is important to keep *C. acnes* in check. Succinic acid, a fatty acid fermentation product of *S. epidermidis*, inhibited *C. acnes* growth. *S. epidermidis* is not commercially available as a probiotic, and researchers are still developing succinic acid as an organic peroxide acne treatment, particularly for those allergic to benzoyl peroxide. Other antimicrobials such as topical benzoyl peroxide and oral tetracyclines (oral antibiotics) will suppress *C. acnes*. In the future, we can expect to see a new generation of probiotics to treat acne in both teenagers and adults.

While oily skin can become troublesome in late middle age, even more common in later life is the opposite problem: The reduction of sebum production, which requires additional lubrication of the skin to protect against pathogens. Sebaceous glands produce less oil as we age. Men experience a minimal decrease after around age eighty, but women gradually start producing less oil after thirty. This leads to dry and cracked skin, particularly around the drier skin sites of the knees and elbows. Dryness may be exacerbated by dysbiosis (the scientific term for a microbial imbalance consisting of less diverse and different microbes), which further disrupts the skin. One way to combat this tendency and thus protect normal microbes is to bathe less often and only in moderate-temperature water. Frequent hot baths can remove moisturizing lipids and oils from the skin and may disrupt the helpful microbes and their defences against pathogens. We'll see later that particular microbial applications, such as the probiotic *Lactobacillus plantarum*, offer promising options to moisten and nourish dry skin.

Ironing Out the Wrinkles

How would you characterize aging skin? Translucency, age spots, and wrinkles probably come to mind. Although microbes turn out to be key players in these signs of age, they can be used to our benefit to improve overall skin appearance. In a twist on personalized medicine, a research group extracted *S. epidermidis* from twenty-one women aged thirty-nine years on average. The scientists grew each person's individual strain in the lab, and then asked the women to add it—or a placebo—to a mixture, and then apply the mixture to their face twice a week for four weeks, always before going to sleep. In the double-blind study (neither the participants nor the scientists knew which group any one participant belonged to), they found that this treatment increased skin lipid content, decreased water evaporation, and markedly improved skin moisture retention while maintaining healthy acidic skin conditions. This suggests that we may soon be able to cultivate our own skin microbes for immediate and future facial rejuvenation. If so, we could "bio-bank" our youthful skin microbes to apply them later in life when wrinkles and dry skin are more of an issue. An intriguingly novel approach to personalized skin care!

There is also some evidence that specific probiotics can be used to target and improve skin appearance. In a separate randomized double-blind, placebo-controlled trial, 110 volunteers aged forty-one to fifty-nine were fed ten billion *Lactobacillus plantarum*, a harmless probiotic, daily for twelve weeks. The investigators found significant increases in skin water content in the face and hands; this hydration boosted the skin so that it appeared healthier with fewer wrinkles. The probiotic significantly decreased skin-wrinkle depth, improved skin gloss, and enhanced skin elasticity and epidermal thickening. The researchers labelled the probiotic a "nutricosmetic agent" for its beneficial effects. *Lactobacillus plantarum* is widely available in commercial probiotic supplements and in fermented foods such as sauerkraut, kimchi, pickles, brined olives, and even sourdough bread.

BOTOX: FROM POISON TO POSH

One of the most unusual ways to control wrinkles is the unconventional use of a deadly bacterial toxin (Toxin A) *Clostridium botulinum*, which lives deep within the soil where there is no oxygen. As you might guess from the name, it produces the deadliest known toxin on the planet, one that causes an often-fatal type of food poisoning called botulism. This organism and its toxin are also famous as a potential bioterrorism agent, since incredibly small quantities of the toxin are so deadly.

Botulism toxin works by blocking neurotransmitter release, and thus paralyzing nerves. Its first real medical use was when very small quantities were injected into muscles that control the eyes, in order to provide a remedy for people who were cross-eyed, and to treat uncontrolled muscle spasms in the face. However, its real fame came when it was discovered that it could minimize wrinkles. Jean and Alastair Carruthers are a Canadian husband and wife team: an ophthalmologist and dermatologist, respectively. In 1987, Jean injected this toxin into the facial muscles of patients to control eyelid twitching and eye spasms. One day, a patient asked why she wasn't injecting her forehead, and Jean replied that her forehead wasn't spasming. The patient replied that when Jean had injected her forehead in the past, her wrinkles had gone away. Naturally this came up at dinner with Jean's dermatologist husband, and they decided to try it the next day on their receptionist who complained of forehead wrinkles. The results were spectacular and Botox was born.

There was remarkable skepticism initially: the concept of injecting the world's most notoriously lethal toxin into a patient for cosmetic purposes was a tough sell. To counter this, Jean injected herself and jokingly says she hasn't frowned since. After conducting controlled studies, Europeans began to use it and North Americans soon followed suit. Botox is now America's number one cosmetic procedure, with six million injections and two billion dollars in sales per year—making it the mainstay of most dermatology practices. However, it is not without its problems: Among the risks, repeated use causes skin sagging and drooping.

We can also potentially use our microbes to address the root cause of most wrinkles and skin aging: sun damage. In a double-blind study of fifty-four people, scientists discovered that feeding participants a particular probiotic (*Lactobacillus johnsonii*) for eight weeks actually decreased sunburn. The fun part is how they studied it: Researchers irradiated one buttock and used the other side as a non-suntanned control. Several trials indicated that probiotics decrease inflammation (which is associated with sun burns, as the skin is damaged), and researchers for this trial suggested that the oral probiotic minimized sunburn by dampening the inflammation and damage associated with UV exposure. The probiotics induced changes in the skin's immune system, which may have boosted its ability to prevent sunburn and promote skin healing.

Scientists hope that bio-engineering of the microbiota will produce a new generation of more effective sunscreens and anti-wrinkle mechanisms. Harvard Medical School researchers have determined how cyanobacteria—an aquatic and photosynthetic bacterium often referred to as blue-green algae—protects itself from UV rays. The bacteria manufacture protective UV-absorbing molecules: mycosporine, mycosporine-like amino acids (MAAs), and scytonemin. Companies already use MAAs to make bio-sunscreens and anti-wrinkle creams, such as Helioguard 365 and Helionori. There is much biotechnological and commercial potential here as we better understand the natural "sunscreen" compounds produced by microbes, as well as specific biological compounds that absorb UV light—although it might be a tough sell to coat oneself in blue-green algae! Keep an eye out for bio-sunscreens that offer protection against premature aging and sun damage by looking for ingredients like *Porphyra umbilicalis* (red algae) extract.

It's All on Your Head

For men, and some women, a tell-tale sign of age is the infamous receding hairline—or worse, a comb-over. While it's easy enough to see with the naked eye whether this process is taking hold of your scalp, we can also now tell whether or not someone is bald just by analyzing a swab of their scalp for microbes. Microbes may be able to help you keep your hair radiant—and on your head—even as you age. There is no data yet for humans, but when probiotic *Lactobacillus reuteri* was fed to mice during the second half of their lives, researchers observed an increase in sebum production, shiny fur, and (as the authors put it) a general "glow of health" within seven days. Preliminary data shows that this probiotic may work by decreasing inflammation through an increase in an anti-inflammatory cytokine (IL-10). The aged male mice eating the probiotic had thicker, more lustrous fur, while the female animals displayed shinier hair. There may be something here we can learn from our furry friends. The probiotic *Lactobacillus reuteri* is readily available in current commercial hair products. The daily dose needs to be large: In the range of one billion to one hundred billion colony-forming units (CFUs are live bacterial cells). *Lactobacillus reuteri*, like most probiotics, can be taken with food, but not with hot beverages (which would cook the bacteria).

We can even potentially take advantage of microbes to treat dandruff, a dermatologic condition that can get worse in older scalps as the skin's structure and function decline. Scientists found that adults with dandruff have higher levels of a yeast-like fungus, *Malassezia restricta*, *Staphylococcus* species, and lower counts of *C. acnes*. The *Malassezia* fungus lives on the scalps of most adults, and it irritates the scalps for some by causing increased growth of skin cells. These extra skin cells die and fall off in pieces: those white and flaky specks visible on scalps and shoulders. It is not proven, but researchers suggest that a microbiota imbalance may have a role in dandruff production. In one double-blind, placebo-controlled randomized trial, participants with moderate to severe dandruff took the oral probiotic *Lactobacillus paracasei* or a placebo. Of those who took the probiotic, 72 percent experienced significantly less dandruff in four to five weeks, compared to 34 percent in the placebo group in the same amount of time. It also had secondary benefits, including reduced scalp erythema (redness or rash), itching, greasiness, and scalp *Malassezia* yeast counts. *Lactobacillus paracasei* is commonly added to probiotic supplements. There are countless websites promoting probiotics to massage into dandruff-infested scalps, and (so far) a formula with live *Lactobacillus paracasei* cultures and at least forty billion CFUs is an intriguing option.

Sweat It Out

A healthy skin microbiome can make us feel attractive in many ways—fewer wrinkles, plumper skin, better hair—but it has one major drawback: It is the culprit behind body odour. Sweat secretions from the body are actually odourless until microbes (mainly *Corynebacterium*) get hold of these secretions and break down proteins and lipids in the secretions into smaller molecules, called volatile organic fatty acids and thioalcohols. These by-products smell but evaporate quickly, hence the body odour that wafts from our armpits.

MYTH: Sweat stinks.

FACT: Sweat is essentially odourless: the source of any smell is the microbes.

Sweating generally becomes less of a problem in later life as the number and activity of sweat glands decreases. However, at any age, the stench of sweat and damp stains can leave you in a pit of embarrassment. We generally control body odour through two main products: antiperspirants, which block sweating and decrease body odour; and deodorants, which don't affect the amount you sweat but decrease body odour. Around 90 percent of Americans use one of these products, resulting in a billion-dollar industry. Antiperspirants contain aluminum salts that dissolve upon application and temporarily block secretion by plugging the sweat glands. They are usually dissolved in ethanol so that they dry faster upon application. Both ethanol and aluminum salts are antibacterial, as are other anti-microbial additives (e.g., triclosan) that help kill the bacteria involved in producing body odour.

One small study examined the effects of deodorants and antiperspirants on the microbiota. Researchers took nine healthy subjects and asked them not to use either product for one month—the time it takes for the underarm skin to turn over. The scientists found distinct microbial communities depending on whether or not a participant used these products. The microbial richness was higher when using these agents, especially the antiperspirants. Ironically, antiperspirants led to an increase in *Corynebacterium*, which is associated with producing body odour.

In the future, we can expect to see more effective products that specifically target odour-producing microbes. A few probiotic deodorants are currently available, though these are mostly organic mixtures with unspecified "shelf stable probiotics" (i.e., live probiotics that are naturally stable at room temperature and do not require refrigeration). DIY recipes combining natural butters and oils with emptied capsules of powdered probiotics abound on the internet. These products likely do not work, as they contain a random assortment of probiotics in varying amounts, not ones specifically involved with the bacteria that cause body odour. As science advances, we can expect to see pharmacy shelves stocked with more precise probiotic deodorants that target specific microbes.

The Future Is Bright

Though it is still early, remarkable strides are being taken to understand the effects of the microbiome on the skin and scalp as we age. Personalized probiotics made from an individual's beneficial microbes show significant promise to both prevent infections and promote skin health. We can imagine a revolutionized approach to skincare in the near future where we analyze our aging skin and add back specific beneficial microbes. In a real twist on personalized medicine, we may even be able to culture and apply our very own "nutricosmetic" microbes to boost skin health and appearance. Similarly, we may take certain microbes from the skin of younger people and apply them in later life as an anti-aging strategy. Biobanking your own personal skin microbes for thirty years may sound futuristic, but we can imagine rejuvenating skincare cocktails that incorporate youthful microbes extracted from younger populations.

Emerging research further suggests that hair growth could be influenced by identifying microbial compounds. Soon we may target balding through microbial interventions that promote the health of hair cells. We may even be able to boost the health and longevity of pigment-producing hair cells to gain more autonomy over whether or not we "go grey."

Scientific advances will filter microbe-friendly skin regimes into our everyday lives. Deodorants of the future may include a precise mixture of microbes that displace those causing body odour. We know that antibiotics wreak havoc on our skin's microbes, especially topical applications. Mixtures of beneficial skin microbiota could become available to apply following antibiotic treatments to help recolonize the skin microbiota in a healthy way. Given the high usage of antibiotics in older populations and a resulting microbe depletion, this is a particularly important therapy to pursue. With evolving knowledge of the vast differences between people's skin microbiota, we can expect probiotic skincare therapies tailored to individuals. We can also imagine a personalized diet that diminishes typical signs of skin aging by modulating the gut microbiota. We have already seen that certain oral probiotics can affect skin health. Diet is a terrific way to modulate microbiota through the gut-skin axis. Microbes are essential to revolutionize skincare and provide a much-needed fresh approach to achieving youthful, healthy, radiant skin.

QUICK TIPS

- **A different way to clean:** Avoid antibacterial soaps whenever possible. Washing hands with regular soap and running water remains one of the best ways to prevent illness and spread infection.
- **Look, don't touch:** Reduce risk of bacterial eye infections by washing hands with soap and water before handling contact lenses. Clean contacts and storage cases thoroughly with fresh disinfecting solution, never water. Leave the empty case open to air dry and replace it at least every three months. If you are an occasional user, do not give bacteria the chance to grow in your static lens case. Never wear lenses stored for more than thirty days without re-disinfecting.
- **Youthful glow:** Certain probiotics, including *Lactobacillus plantarum*, show significant promise for decreasing skin wrinkles and promoting general skin health. Consider incorporating *Lactobacillus plantarum* into your diet through a probiotic supplement and fermented foods (e.g., sauerkraut, kimchi, pickles, brined olives).
- **Sun safe:** The easiest and least expensive way to keep skin healthy and youthful is to stay out of the sun. Emerging bio-sunscreens, such as Helioguard 365 and Helionori, use cyanobacteria UVA filters to naturally protect against the sun; they may be a better alternative to traditional sunscreens, which negatively disrupt the skin microbiota. After any extended sun exposure, consider taking the probiotic *Lactobacillus johnsonii* to help decrease inflammation and sunburn.

3

Mind Your Microbes: Microbes and the Brain

The brain is key to your sense of self. It provides you with a unique identity, different from every other human; and yet our brains are what bring us all together, as the most intellectually advanced species. We didn't always have such a big head about the importance of our brains to the human species. Historically, it was thought that the gut could be a source of reasoning, hence the common expressions *gut feeling* and *go with your gut.* Over the past 150 years, scientists came to understand that we operate with a much more top-down approach, with the brain controlling every part of the body. The process seems simple: Upon receiving signals from the brain via connecting nerves, the body obeys. Not so fast. Nerves are not doing all the work, it seems; rather the microbes in the gut have something to say about all this, too. They are in constant communication with the brain, either directly or through other mechanisms that alter the brain. This connection is turning into one of the most fascinating and complex areas of microbiome research.

We spoke to Dr. Brian MacVicar, a distinguished neuroscientist at the University of British Columbia and co-director of the Djavad Mowafaghian Centre for Brain Health. He studies glial cells, which surround and interact with our neurons to control everyday behaviour and internal functioning. His research led him to a transformation in thinking about glial cells—specifically the microglia, one of two classes of glial cells. As he told us, "We knew that the microglia were involved in infections and reactions, that they were part of the immune system. We thought they were just sitting there, and that they slowly transformed with diseases after several days." But using new technology, he and his team saw that these microglia were moving all the time, actively checking on the health of brain tissue. This groundbreaking discovery was the first step that led him to consider microbial connections to the brain.

When he first suggested pursuing gut-brain research at scientific meetings in 2012, Dr. MacVicar faced widespread skepticism and doubt. But since then, the entire neurobiology field and research climate have changed. An influential paper from Germany showed direct communication between the brain and microbiome. The researchers studied germ-free mice, which lack a microbiome, and discovered that the microglia cells in their brains were different from mice with microbes. This was a pivotal moment that demonstrated interaction between the brain and microbiome. As Dr. MacVicar explains, the discovery has huge implications for dementia, Alzheimer's, Parkinson's, multiple sclerosis, and other brain diseases.

MYTH: The brain controls every part of the body and tells it exactly what to do.

FACT: There is two-way communication between the gut and the brain.

Microbiome-brain research, he says, is now the "flavour of the week—or hopefully decade; maybe even the flavour of the next generation!" It is so trendy that the number of speculative articles about it in every major neurology journal is far greater than the handful of actual data-based, peer-reviewed papers currently published. Dr. MacVicar acknowledges that while there is much enthusiasm, lots of work and questions lie ahead for scientists to address: "We know that there is definite communication between the microbiome and brain, but how *exactly* do they communicate? We don't have a clue about the mechanisms yet." Understanding this process would represent a huge step in society's ability to leverage the microbiota and promote brain health.

The Gut–Brain Axis

Control of the gut's functions specifically occurs via the brain, but the communication is reciprocal and happens along what is often called the gut-brain axis, a pathway along the vagus nerve. This major throughway in the body runs down from the brain, winds through the nerves in the gut ("vagus" means *wanderer* in Latin), and hooks up the gut neurons, or enteric neurons, with signals from the brain. The number of neurons in the gut is second only to the number in the brain, so together these two systems form a powerful communication hub.

So, how might the gut microbiota affect the brain? Increasing data indicates that microbes use three major routes to "talk" to the brain. The first involves the main conduit on the gut-brain axis, the vagus nerve. By influencing gut neurons directly hooked up to the vagus nerve, gut signals, which the microbes can also influence directly, can be communicated to the brain. The second route involves microbes making chemicals (neurotransmitters and hormones) that can signal the brain through other nerve networks. The third route involves influencing the immune system, which interfaces directly with the nervous system throughout our entire bodies. And as we'll see throughout this book, microbes love to tweak the immune system.

There are two major barriers that physically separate the gut and the brain and complicate communication. The first is the intestinal wall, a tube passing through our entire body that keeps undigested food and microbes confined and transports nutrients from the gut into the body to use for energy. Although it sounds counterintuitive, what's inside the intestinal tract is actually considered *outside* our bodies, since this tract it really just a tube that runs through us. Several animal studies have shown that holes in this tube—a.k.a. a leaky gut—have a major impact on the brain. If microbial products leak into the body, it immediately responds with inflammation. Think of it as tactical military strikes trying to block invading microbes from entering the body—a line of defence against incoming infections. As in a real battle, inflammation does a lot of damage to our own body, even as it seeks to protect it. This can result in brain problems that we'll explore later.

The second major barrier is aptly named the blood-brain barrier, or BBB. It is formed by the special endothelial cells that line blood vessels in the brain. So although there are lots of blood vessels running throughout the brain, their contents are kept separate (except for specific nutrients and oxygen that are transported across). The BBB is incredibly efficient at keeping most things out of the brain; this is why delivering chemicals/medicine to the brain via drugs is so difficult. Several animal studies have shown that microbes are needed to fully develop this barrier. Germ-free animals have a more permeable BBB because microbes influence the expression of key proteins that form seals between endothelial cells, gluing the cellular joints together to make the barrier. Amazingly, if these germ-free adult animals with leaky BBBs are then colonized by intestinal microbes, their BBB permeability goes down (i.e., the barrier reseals itself).

Describing all the pathways and mechanisms that microbes use to influence brain activity is a large, complex, and still confusing proposition. So instead of trying to tackle the work of future scientists, we will focus the rest of this chapter on key brain issues associated with aging—there are many! And we will explore how the microbiota might influence these major geriatric diseases.

What we have found is really quite amazing. While we are still determining the extent to which we can use this information to improve cognitive health, the field is a fast-moving one. We are optimistic that future findings, even just a few years from now, will have a very big impact on lifelong mental well-being and healthy aging. We also include some useful tips and habits for what you can do *now* to directly influence your microbiota.

The Aging Mind

We've all misplaced keys, blanked on the name of a person or place, or forgotten a phone number. When we're young, we don't often pay much attention to these lapses; however, as we grow older we increasingly worry about what they could mean. Close to four in ten people over the age of sixty-five experience some form of memory loss, known as "age associated memory impairment." There is no underlying medical condition for this; it is considered a normal part of aging, since getting older means all parts of our bodies—including the brain—can slow down. When starting to experience this, many aging people are most afraid of dementia, which involves the loss of mental function (namely thinking, memory, and reasoning): i.e., losing one's unique source of personhood. Either in our own lives or in movies, we've all seen victims of late-stage dementia—an elderly woman who no longer recognizes her own children or an older man yelling inappropriately for no reason. One new case of dementia occurs every three seconds in the world, and naturally we fear falling victim to its gradual and relentless march to memory loss, confusion, irritability, wandering, and even the inability to speak or recognize common things.

But people frequently mix up normal age-related memory loss and dementia. Dementia involves cognitive impairment that is severe enough to interfere with a person's daily functioning. It is *not* a normal part of aging (see Table, page 40), and it is not itself a disease; rather, it comprises a group of symptoms of mental loss caused by a number of different diseases and conditions such as Alzheimer's disease, Lewy body disease, cerebrovascular disease, and other cognitive impairments. Nor does old age *cause* dementia. Age is, however, the biggest risk factor. In other words: Dementia can happen to anyone, but it is more common after the age of sixty-five (4 in 1,000 people aged sixty to sixty-four are diagnosed with dementia, in comparison

MYTH: Intelligence declines with age.

FACT: Crystallized intelligence—knowledge and experience accumulated over time—remains stable as we age. However, fluid intelligence, the ability to think logically and solve problems in novel situations not based on experience or education, tends to decline.

to 105 in 1,000 people aged ninety and older). It costs billions of dollars for treatment and care; if global dementia care were a country, it would be the eighteenth-largest economy in the world!

MEMORY LOSS AS WE AGE: WHAT IS NORMAL?

Normal age-related memory changes	*Symptoms that may indicate dementia*
Able to live independently and participate in normal activities despite occasional memory lapses	Difficulty with everyday simple tasks such as paying bills, bathing, dressing appropriately; forgetting how to do things that have been done many times
Can remember and describe moments of forgetfulness	Cannot remember or describe even significant moments of forgetfulness
May pause to remember directions, but does not get lost in familiar places	Gets lost in familiar places; unable to follow directions
Sometimes has difficulty finding the right word, but has no trouble holding a conversation	Frequently forgets, misuses, or garbles words; regularly repeats lines and stories in the same conversation
Maintains judgment and decision-making capabilities	Trouble with choices and sound judgment; behaves in socially inappropriate ways

Source: Smith, M., Robinson, L, Segal, R. 2018. Age-Related Memory Loss (helpguide.org/articles/alzheimers-dementia-aging/age-related-memory-loss.htm).

We have all had the experience of being called on by a teacher in school and not knowing the answer. We felt our collar tightening, voice faltering, palms sweating, and face turning red. A person with Alzheimer's disease sits every day in a giant classroom not knowing the right answers (as described in *The Best Friends' Approach to Alzheimer's Care* by Virginia Bell and David Troxel). Their brain is like an iceberg: there are some days when sections melt away and are lost, and other days when it holds its shape. People with the disease often feel confused, embarrassed, frustrated, angry, lonely, and sad.

Alzheimer's disease is an irreversible illness that slowly and progressively damages nerve cells in the brain. Most people's first symptom is forgetfulness—which seems harmless—but symptoms gradually worsen as more brain cells are destroyed, to the point where some people can no longer form coherent words or feed themselves.

The disease is increasingly common: Alzheimer's represented the sixth leading cause of death in United States in 2016 according to the CDC, and is expected to increase threefold by 2050. Sadly, it is the only disease on the CDC's list for which there is no way to prevent or cure its progression. Someone with Alzheimer's disease tends to live four to eight years after diagnosis, though some may live as long as twenty years.

What do microbes have to do with dementia and Alzheimer's? Although preliminary research is still being conducted, there are several indications that the microbiota may have an impact—direct or indirect—on these brain diseases. The rate of Alzheimer's is much higher in developed countries—a hint that increased hygiene and the Western diet, and their microbial side effects, could be contributing factors. In addition, urban areas have up to ten times more cases of Alzheimer's than rural areas where all sorts of good microbe-rich stuff lives (more on the environment in Chapter 13).

High levels of circulating sugars are not healthy for the body and brain. As we'll see later, gut microbes play a major role in both diabetes and blood sugar levels. Diabetics, who often have high levels of sugar in their blood, are twice as likely to develop Alzheimer's disease; and, indeed, Alzheimer's is unofficially sometimes called "type 3 diabetes."

Inflammation is a normal response of your body's immune system to trauma and harmful bacteria trying to invade your body. It acts as both a friend and foe because, while inflammation is an essential line of defence against acute threats, long-term, nonspecific inflammation (lasting months to years) is harmful to your body. It severely disturbs your microbes and features in a wide range of chronic conditions (more on this in Chapter 11). Chronic low-grade inflammation is an especially significant risk factor for Alzheimer's disease. Alzheimer's patients have up to three times the amount of lipopolysaccharide—or LPS, a predominant surface molecule that is found in many bacteria—circulating in their blood than normal controls. LPS and other bacterial molecules trigger inflammation in our bodies, our immune system's 911 response to these bacterial signals, thinking a nasty pathogen is invading. While useful for controlling infections, chronic inflammation is bad since it causes significant cell and tissue damage—think of arthritis. Both dementia and Alzheimer's patients have increases in many of the normal markers of inflammation—such as inflammatory cytokines, which indicate inflammation is occurring—circulating throughout the body.

We know that the gut barrier and the blood-brain barrier become more permeable with age, making them less capable of keeping less beneficial microbial products out. When they seep across these barriers, we experience low-grade chronic inflammation. Gut microbes in Alzheimer's patients are different than in controls without the disease, and this dysbiosis may contribute to the increased inflammation, including increased LPS release. We also see Alzheimer's patients with increased permeability in the blood-brain barrier, which would allow inflammatory microbial products to cross into the brain and trigger more inflammation.

Certain cells in the brain respond to inflammation by increasing the production of a molecule called amyloid beta. This forms the tangle of intertwined protein that makes up the characteristic spider webs seen in the brains of Alzheimer's patients—plaque that builds up between nerve cells. This neuroinflammation also causes other brain damage, contributing to overall learning and memory impairment and cognitive decline. Multiple studies have shown that taking anti-inflammatories for more than two years to decrease inflammation can significantly decrease the risk of both Alzheimer's and Parkinson's diseases.

Perhaps the strongest evidence of a microbial connection with Alzheimer's comes from a traditional Alzheimer's model using mice. In this model, diseased mice have a very different microbiota than normal control animals. Recently researchers found that if these animals were raised without microbes, they had a remarkable reduction in brain pathology of over 70 percent, and also much less neuroinflammation. When these microbially sterile mice were recolonized with feces from diseased mice, they had a much higher rate of pathology than if they were recolonized with feces from normal mice.

In the Alzheimer's field, controversy rages over the claim that certain infections are associated with increased risk of Alzheimer's disease. These are infections caused by a virus—Herpes Simplex Virus (HSV-1)—or by certain bacteria—*Chlamydia pneumoniae*, which causes pneumonia and eye infections, and *Borrelia*, which is associated with Lyme disease. Although this connection has not been proven, people with a genetic susceptibility to Alzheimer's have more frequent infections from these organisms. Again, the concept is the same: increased brain inflammation from infection may trigger increased amyloid production and plaque formations, thereby triggering Alzheimer's.

An Ounce of Prevention

Identifying the exact ways microbes may influence the brain is a big step forward, but the bigger question remains: How can we use this information to keep our brains healthy? Although there is currently no cure for Alzheimer's or other forms of dementia, treatments are available to alleviate symptoms. Medications (such as donepezil, galantamine, memantine, rivastigmine) slow the progression of neurodegenerative diseases, but you can also do a lot by proactively modifying your lifestyle and diet to minimize the risk for dementia and reduce your symptoms if you already have a condition.

First, it is important to stay away from tobacco and smoking, which deprive your brain cells of oxygen and important nutrients. Second, limit alcohol intake to recommended limits, including a glass of red wine a day (see later in this chapter). Aside from all the other dangers of intoxication, binge drinking and heavy sustained drinking can cause alcohol-related brain damage. Third, eating a healthy and balanced diet helps to keep your brain from "rusting." Even increased coffee consumption is associated with decreased risk of Alzheimer's. Although the mechanism behind it is not known, coffee is known to move the gut microbe composition away from being toxin-producing and inflammatory.

THE MIND DIET: MIND WHAT YOU EAT

Given the links between diet and the microbiota, there is growing and compelling evidence that diet affects the health of our brains via the microbiome. Two randomized trials of the Mediterranean diet and DASH (Dietary Approaches to Stop Hypertension) found that following these diets had protective effects against cognitive decline. Researchers at Rush University Medical Center in Chicago took this information and developed a new diet specifically tailored to protect the brain: the MIND diet (Mediterranean-DASH Intervention for Neurodegenerative Delay; see rush.edu/news/diet-may-help-prevent-alzheimers). MIND includes mainly natural plant-based foods and limits meat and high saturated fat.

The researchers purposefully included foods in the diet that have been shown by compelling evidence to protect the brain against dementia. For example, a number of studies have observed that people who consume high amounts of vegetables, especially green leafy vegetables, have slower declines in cognitive abilities. One large study using animals showed that berries may also protect the brain; in another study, rats who were fed grapeseed extract (grapes are scientifically considered a berry) digested the micronutrients and produced phenolic acids via their intestinal microbiota. This increased the accumulation of two key acids in the brain—3-hydroxybenzoic acid and 3-(3′-hydroxyphenyl) propionic acid—that are known to protect against Alzheimer's disease.

More research is needed to confirm these results, but the MIND diet clearly shows a promising approach to protection against cognitive decline. In a 2015 study, the MIND diet lowered the risk for Alzheimer's disease by 53 percent in participants who carefully adhered to it (regardless of other factors such as lifestyle and heart health). Even modest adherence to the diet reduced Alzheimer's risk by 35 percent. Besides, the MIND diet has components that are already proven to reduce risk for hypertension, heart disease, and stroke.

As you can see in the chart below, the diet is fairly easy to follow. Have a green salad and another vegetable every day, and snack on nuts. Enjoy a glass of wine once a day to keep your taste buds and brain happy. Here is a list of fifteen foods to include and to avoid for brain-boosting meals.

THE MIND DIET

Include These

- Green leafy vegetables: every day
- Other vegetables: at least once per day
- Whole grains: three times per day
- Nuts: every day
- Wine: one glass per day
- Beans: every other day
- Berries: at least twice per week, especially blueberries and strawberries
- Poultry: at least twice per week
- Fish: at least once per week
- Olive oil

Limit These

- Red meats: fewer than four servings per week
- Butter and stick margarine: less than one tablespoon per day
- Cheese: less than one serving per week
- Fried or fast food: less than one serving per week
- Refined sugars and carbohydrates (e.g., pastries and sweets): limit as much as possible (e.g. special occasions)

The beauty of the MIND diet is that it benefits your brain even if you are not following it to the letter, so enjoy what you like most from this list and you can still boost your microbiota.

Along with a healthy diet, it's essential to exercise your body regularly throughout your life. When you get your heart pumping, blood flow to the brain increases, providing additional nourishment. Regular physical activity reduces risk factors for dementia including high blood pressure, diabetes, and high cholesterol. Exercise also affects microbes, as we will see in Chapter 12. A growing and convincing body of evidence shows that regular aerobic exercise not only helps protect your brain from Alzheimer's and dementia—it can also improve your quality of life if you already have the disease. Results from several randomized controlled trials (the gold standard of science experiments) showed that people with Alzheimer's on aerobic exercise programs achieved higher cognitive scores, improved memory status, and felt more alert and organized and less anxious, irritable, and depressed. To boost results—and to find that much-needed motivation to get out the door—consider physical activities that are also mentally and socially engaging. Walk or jog with a friend, join a group exercise class, volunteer in a local animal shelter or garden—whatever you do, make sure it's something that you enjoy.

Parkinson's Disease

Brett has been a serious classical clarinet player since school days, but during an early forties mid-life crisis he decided to become a "cool" jazz sax player. He began playing tenor saxophone and other woodwinds in what became known as the Oscar Hicks Jazz Sextet. The bass player in the group, Rod, is an outstanding musician who breathes, sleeps, and eats jazz. He patiently transformed the group from neophyte, classically trained jazzers (read: improvisation was not on their theory syllabi) into a reputable jazz band with actual paying gigs. The close-knit group is still together twelve years later and actively plays in Vancouver at various venues, rehearsing and drinking good single malt whisky every Wednesday night. Not bad for a mid-life crisis, right?

But about two years ago, Rod, then in his early sixties, noticed that his left hand shook and twitched when it was resting. His gait also became a bit more rigid, and his face seemed tighter when he smiled. When he finally saw a neurologist, he was diagnosed with Parkinson's disease. In addition to the tremor, he struggled with fine motor control. The disease made his hand "feel like a club," and playing bass became very difficult. "The subtlety and nuance required for smooth technique disappears and makes playing feel lumpy, primitive, and no fun at all," he said. Rod had to quit playing bass.

> **MYTH:** If you are diagnosed with Parkinson's disease, there is not much you can do.
>
> **FACT:** A Parkinson's diagnosis does not necessarily mean a shortened life, and there are effective ways to control the symptoms such as medication, exercise, and diet.

Luckily, Parkinson's is a slow-acting disease. Although not curable, there are ways to control the symptoms so that "Parkies" (a self-proclaimed label) can function in society for many years. Rod used to eat a lot of meat and would often buy fruit only to let it spoil before he got to it. Since his diagnosis, he has become much more aware of his diet and its potential impact on the microbiota. He makes fruit smoothies with raspberries, strawberries, bananas, and mango for breakfast; and in general, he has decreased his meal sizes and meat portions. Rod previously had a sedentary lifestyle, but now he walks his dog twice a day and rides his bike more often. He also takes Sinemet (the brand name for Carbidopa and Levodopa) three times daily. Sinemet is a medication that the body converts into dopamine—the neurotransmitter that is lacking in Parkinson's and affects motor control. His symptoms are under control. As a Canadian, he's also able to experiment with high-CBD medical marijuana, which is prescribed by his doctor and which he feels helps to reduce his tremor, elevate his mood, and enhance his sleep. But as he says: *That's another story for another time.* To everyone's delight, Rod is back playing strongly and reunited with everyone for gigs and practices (and whisky tasting). There is an expression in jazz that seems appropriate as an end to Rod's story: "Play and solo like there is no tomorrow."

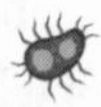

Parkinson's disease is the second most common neurodegenerative disease (after Alzheimer's), and affects 1 to 2 percent of the population over the age of sixty-five. It is a neurological disease with symptoms that include those Rod experienced: tremors, rigidity, slow movement, and facial stiffness. Although the symptoms first present as a neurodegenerative disease, Parkinson's may start twenty to thirty years earlier in the gut. The two most common early clues of Parkinson's are constipation in the gut and loss of sense of smell, both of which hint at microbe involvement. People usually live for many years after onset (it generally doesn't affect lifespan), and the motor symptoms can be controlled to a fair extent by dopamine for several years. High profile people affected by the disease such as

MYTH: Symptoms of Parkinson's disease start in the brain.

FACT: The first symptoms often begin in the gut with constipation, and sometimes in the nasal cavity with loss of smell.

Mohammad Ali and Michael J. Fox have increased public awareness, and are examples of how it's possible to continue most aspects of one's daily routine for years before the symptoms become too severe.

A protein called alpha-synuclein plays a key role in Parkinson's disease as the *substantia nigra* (a specific area of the brain that helps coordinate muscle control by producing the neurotransmitter dopamine) is slowly destroyed. The gut also expresses alpha-synuclein, which can become misfolded in the gut—think of a ball of knitting yarn that becomes tangled into a knot. Scientists think that these misfolds have a domino-like effect on misfolding synuclein moving up the vagus nerve into the brain, causing plaques to build up similar to those seen in Alzheimer's, and ultimately leading to the destruction of the dopamine-producing cells. Think back to that ball of yarn: If you just tug and pull at the knot haphazardly, the knot gets bigger and bigger until you have a giant tangled mess.

As with Alzheimer's disease, there are currently no ways of reversing the damage to the *substantia nigra*. Researchers have tried to use stem cells but were not successful. Healthy eating, exercise, and diet modifications (see the box on the MIND diet) seem to slow the progression of Parkinson's to some extent, and dopamine can be used to help manage symptoms.

With the realization that the gut and perhaps, due to loss of smell, the oral-nasal route, are involved early in Parkinson's, attention has recently shifted to exploring how our body's microbiota may contribute to this disease. There is no causal data yet, but there are several smoking guns hinting at gut and microbe involvement, which we'll explore here. Collectively, all the recent data points to Parkinson's disease being associated with gut microbes *before* brain symptoms are seen. This constitutes a paradigm shift in thinking about the disease.

Remember the vagus nerve, or the main "telephone line" connecting the gut and brain? One line of evidence comes from a medical procedure seemingly unrelated to Parkinson's: Cutting the vagus nerve to help with peptic (stomach) ulcer pain. In a fascinating study conducted between 1977 and 1995, researchers in Denmark tracked over five thousand people whose vagus nerve was cut. Five years later, researchers examined the group's Parkinson's rates. What they found was startling: Patients who had the nerve severed completely had less risk of Parkinson's than those who either had the nerve partially severed or had it left completely intact. A very recent Swedish study involving close to ten thousand patients who had this nerve cut supports the Danish finding, showing that complete, but not partial, cutting of the vagus nerve had a protective effect against Parkinson's. It could be that severing this nerve blocks misfolded synuclein from reaching the brain (this has actually been shown in rats), or that gut signals, including those from microbes, are blocked as a result.

A recent animal study provides even more compelling evidence that Parkinson's may start in the gut. A model of this disease used mice that overexpress alpha-synuclein, and therefore develop motor disabilities and synuclein aggregates. Researchers showed that if these mice were raised without microbes, they showed significantly fewer motor disabilities and alpha-synuclein aggregates. If researchers added short-chain fatty acids—which are produced by gut microbes—to these germ-free mice, the mice had more disease symptoms. The most remarkable part of this study is that researchers placed feces from Parkinson's patients in the mice, and the feces from the Parkinson's patients increased symptoms of the disease. Feces from the controls did not increase disease symptoms.

A common theme emerging in our discussion about aging and its associated diseases is increased chronic inflammation in the body—inflammaging. Inflammatory microbial products such as LPS—lipopolysaccharide, a molecule found on the surface of many bacteria—are a key component of this effect. As with Alzheimer's, Parkinson's patients have been shown to have increased gut permeability. The resulting leakage of microbial products into the body triggers inflammation in the gut as well as in the rest of the body. It is currently thought that this inflammation may occur many years before any appearance of brain symptoms and may lead to increased misfolding of alpha synuclein throughout the enteric nerve system, including the vagus nerve and moving all the way up to the brain.

So what starts this process of inflammation? We have clues that the intestinal microbiota may play a significant role. The general idea emerging from various studies is that gut microbes in Parkinson's patients have shifted to include more microbes that produce the inflammatory molecule LPS, and fewer microbes that dampen inflammation. This causes increased inflammation, which is associated with Parkinson's.

A small number of human studies show that the microbiota in Parkinson's patients is different than in controls who don't have Parkinson's. One study found a staggering 77 percent decrease in a bacterium called *Prevotella* among Parkinson's patients. They found that just by measuring the decrease in this bacterium they could distinguish between patients and controls. This is important, as there is currently no good biomarker to identify Parkinson's early, meaning doctors normally have to wait until the neurological symptoms appear in order to diagnose. *Prevotella* has been associated with promoting a healthy gut by increasing short-chain fatty acids that decrease inflammation and by making vitamins. Researchers also know that the prevalence of this bug increases in people who follow diets rich in fibre, fruit, and vegetables. Sound familiar? The study found that bacteria that are thought to be more inflammatory (*Enterobacteriaceae*, which produce LPS) were associated with patients who had increased severity of motor defects. Similar differences in these microbes have been reported for another neurological disorder: autism. A further study found that Parkinson's patients had lower levels of "good" microbes associated with anti-inflammation, and higher levels of "bad" microbes associated with inflammation.

Constipation is another risk factor for Parkinson's, with up to 80 percent of diagnosed patients reporting this gastrointestinal issue. The reason may have to do with the fact that due to slower transit time, constipation can lead to alpha-synuclein accumulation in the gut—which, as we know, causes it to misfold and trigger the disease like that knotted ball of yarn. Similarly, constipation could precede the microbial dysbiosis; or the microbial imbalance could trigger it, resulting in increased intestinal permeability and inflammation. About half of all Parkinson's patients suffer from depression. As we will discuss in the next section, depression is increasingly associated with changes in the microbiome. Perhaps dysbiosis in the gut may contribute to such depression?

There are two rather surprising lifestyle traits associated with reduced risk for Parkinson's disease. The first is smoking, which has been documented to decrease the risk by 36 to 50 percent; the second is regular coffee consumption, which decreases risk by about one-third. There are lots of ideas, and of course no concrete answers, as to why this is the case. One involves the gut. As any morning coffee drinker knows, coffee stimulates bowel movements—within four minutes according to studies. Perhaps coffee keeps one regular, which deters the constipation so often associated with Parkinson's. Coffee also has antioxidants that could decrease inflammation, and thereby protect the misfolding of alpha synuclein.

Mental Health

Depression is one of the most frequent causes of emotional suffering in old age, yet it often goes undiagnosed and is inadequately treated. Many people think that as health problems occur and loved ones pass away, it is inevitable for elderly people to become depressed. Not true! According to the experts, older adults who live independently (non-institutionalized) have a low rate of clinical depression: 1 to 5 percent in large-scale US and international studies, in comparison to 6.7 percent in the general American population. However, depression rates *do* greatly increase, to over 30 percent, with hospitalization and medical illness (patients with stroke, heart attack, or cancer have rates over 40 percent). As many as 50 percent of nursing home residents are depressed. More than half of depressed older adults have their first episode after the age of sixty. Stress and anxiety go hand in hand with depression—as in younger adults, anxiety typically precedes depression and at least 50 percent of depressed older adults have anxiety disorders. However, in other respects, the condition often looks different in older adults compared to younger adults; for example, depressed older adults do not generally report sadness or worthlessness. Instead, their symptoms may include insomnia, apathy, decreased energy, loss of appetite, agitation, poor memory and concentration, and physical aches and pains.

MYTH: Depression is a normal part of aging due to illness and personal loss.

FACT: Depression is not caused by aging and should be treated.

No one knows exactly what causes depression, which can make it difficult to diagnose and treat. Making the situation worse, late-life depression (especially in men) is often undiagnosed or misdiagnosed, including being confused with the effects of multiple health conditions and the medicines used to treat them. Undiagnosed depression can result in severe harm—even suicide. The latter is especially troubling, as suicide rates are almost twice as high in elderly people—white men aged eighty-five and older have the highest rate of suicide in the US. Just as depression has no single cause, there is no one treatment that works for everyone. Typical treatments involve some combination of therapy, medication, and lifestyle interventions.

As we've seen above, the gut microbiota appears to have a large influence on brain function, much more than we ever imagined. Some compelling recent animal studies have shown that behaviour, anxiety, stress, and depression are all linked to the gut microbiota. There is quite strong data that early-life microbiota play a major role in these mood disorders. More evidence is also emerging concerning adult microbiota; for example, germ-free mice have very different behaviours—they become more daring (like many "prey" animals low on the food chain, mice are normally pretty cautious) than germ-filled controls. Even more remarkable: transfers of feces from stressed, anxious, or depressed rodents into normal laboratory animals cause the control group to exhibit stress, anxiety, or depressive behavior. The fact that these behavioural issues can be altered by simply modifying gut microbiota has profound implications for the potential of altering these diseases in humans. Many indications increasingly support this.

Two small studies recently pointed to a conclusion that people with mental health problems have an altered gut microbiota. One looked at the fecal microbiota in thirty-four depressed patients, compared to seventeen controls, and found significant differences: Depressed patients had increased levels of the "inflammatory" microbes, such as Proteobacter and bacteroidetes that make LPS, and decreased levels of less-inflammatory *Firmicutes*, a group of bacteria including some that produce anti-inflammatory effects. The scientists also found that levels of a small molecule called isovaleric acid, which is produced by microbes, were increased in depressed patients. Isovaleric acid—a compound that chemically resembles the neurotransmitter GABA, which works to balance mood—can cross the blood-brain barrier, where it can affect neurotransmitter function by competing with the GABA receptor. Think of this as a microbial interference with the chemical messaging sent from our brain throughout our body. In the second study, which was of a similar size, a different group of researchers found increases in *Bacteroidales* and decreases in *Lachnospiraceae*—but the correlation between species was "complex" (a scientific term meaning: "we just don't know what it means yet . . ."). Much larger studies involving many more people are needed to make sense of these potential differences and to draw conclusions, but the concept of being able to detect "depressing microbes" and their by-products in the future opens new ways of thinking about these mood disorders.

By far the most compelling line of evidence that gut microbes may affect depression comes from studies on antibiotic use, the role of which in the health of our microbiota we'll discuss later in detail. A massive retrospective study looked at antibiotic usage in over two hundred thousand people with depression in the UK. The researchers followed people who had been treated with a single course of an antibiotic over a year ago, and found that those treated with any of the seven different classes of antibiotics had an increased risk for depression. Those who were treated with more than five courses of penicillin had a 50 percent increase in the risk for depression. They also found that antibiotics increased the risk for anxiety. The researchers found no correlation with anti-virals, and only a small one with one course of anti-fungals, but not with repeated courses. This all points to the involvement of bacteria in depression, but not viruses or fungi. It also serves as a reminder that antibiotics can be harmful in ways we're only just beginning to understand. Overuse of them can affect not only our physical health, but our mental health, including depression, anxiety, and stress, as well.

MYTH: Effects of stress occur only in the brain.

FACT: Stress causes changes across the entire body including muscle tension, fatigue, upset stomach, and increased gut permeability.

Chronic stress can have a number of negative health effects, ranging from insomnia to weight gain, increased risk of heart disease, and disrupted immune and digestive systems. Stress increases gut permeability; scientists have measured this by stressing individuals through public speaking and cold-water treatments. A study of seventy-three Norwegian soldiers who, on strict rations and carrying a forty-five-kilogram (one hundred–pound) pack, cross-country skied fifty-one kilometres over four days, showed that stress does indeed increase intestinal permeability and inflammation, as well as producing significant changes in the intestinal microbes. As we have seen, increased gut permeability leads to an increase in the leakage of inflammatory bacterial products such as LPS into the body. Anti-LPS antibodies (which protect the body) are higher in number in depressed patients, suggesting increased LPS leakage. Further evidence for this comes from diet studies, where we see that unhealthy diets significantly correlate with chronic inflammation and depression. Strict adherence to the Mediterranean diet correlates with decreased risk of depression.

One of the ways probiotics are thought to work is by decreasing gut permeability, which may impact mood. Extensive studies in animals show that probiotics can indeed decrease depression, anxiety, and stress. There are similar models for neurodegenerative and mood disorders. However, the data in humans is only just starting to emerge, and we need more studies on large numbers of people with diverse health backgrounds. In one study of fifty-five healthy people, researchers administered either the probiotics *Lactobacillus helveticus* and *Bifidobacter longum* for thirty days or a placebo control. When the subjects were subsequently evaluated for anxiety, depression, stress, and coping mechanisms, they found that the people on probiotics fared much better. A similar study of twenty healthy people who were given a mix of probiotics showed that probiotics reduced negative thoughts associated with sad moods. But remember, these were healthy individuals to begin with. There are potentially many ways probiotics could affect the brain, including making chemicals that have direct or indirect effects on brain health (including inhibiting neuroinflammation). The above studies suggest that further investigations are certainly worth trying.

Strokes

Imagine being trapped inside your own body—paralyzed physically but still mentally "you." This is what happens in a stroke, one of the most feared health problems among older people, along with dementia. And rightly so: Strokes are the fifth leading cause of death in the United States and remain the leading cause of adult disabilities.

A stroke occurs when blood flow to the brain is blocked. The lack of blood supply causes cell death within the brain and subsequent numbness, paralysis, loss of speech, and other symptoms depending on which area of the brain is affected. Early treatment and therapy can minimize brain damage and restore function, but there is usually some residual damage to brain cells and less than complete recovery.

There are two general types of strokes. The most common are ischemic strokes, which cause blockage within blood vessels that feed the brain.

Hemorrhagic strokes are less common but more fatal; they result from the rupture of a vessel in the brain and subsequent bleeding and are often caused by high blood pressure. Fatty cholesterol deposits in blood vessels are a major contributor to both kinds of strokes, as they restrict blood flow. Controllable risk factors—including high blood pressure, diabetes, obesity, high cholesterol, excessive alcohol, and smoking—account for half the risk of stroke. The uncontrollable factors that account for the other half include age (older than sixty-five), gender (men have more strokes, although women tend to have more deadly strokes), race (African Americans are at increased risk), and family history.

MYTH: There is nothing you can do to prevent a stroke.

FACT: Up to 80 percent of strokes are preventable. There is a lot one can do, including managing blood pressure and cholesterol, maintaining an appropriate weight, and eating a heart-healthy diet full of fiber.

Just like blood supply to the heart, blood flow to the brain is critical for all functioning, so any cutting off of the blood supply will have other major implications for the body besides strokes. The health of the arteries that supply the blood is critical for both the brain and heart—and the rest of the body. Plaque—consisting of cholesterol and fats—can build up and plug the arteries, which causes cardiovascular disease (CVD), and blocked arteries directly affect the brain through strokes. We'll discuss heart attacks, strokes, CVD, and the microbiota in much more detail in Chapter 8.

The microbiota has major effects on blood vessel health, which affects CVD and, by extension, strokes. When you eat red meat, the microbiota in the meat itself convert components of meat into specific compounds (called trimethylamine, or TMA); the liver then converts these compounds into a derivative compound (trimethylamine N-oxide, or TMAO) which then causes plaque accumulation and CVD. Without red meat and its accompanying microbes, these compounds are not made, which drastically reduces the incidence of CVD and stroke. Eating red meat enriches the microbiota that produce these problematic compounds, and the differences in microbiota (and TMAO levels) can, in a crude sense, be used to provide an indication for risk of stroke.

There are easy ways to decrease the risk of CVD, including reducing red meat consumption and increasing fibre intake. The latter seems to be an especially good way of decreasing stroke risk; we can look to studies involving vegans and vegetarians—who consume a lot of plant-based fibre—and germ-free mice, which have no microbes. All have little to no CVD. This very likely has something to do with the microbiota, since fibre has profound effects on microbiota composition. A recent meta study (a compilation of the findings of many studies) found an inverse correlation of stroke risk and fibre intake. Increasing dietary fibre by just seven grams a day lowered the stroke risk by 7 percent. To put this in perspective, the average fibre intake in the USA for women is thirteen grams a day (recommended is twenty-one to twenty-five grams a day), and for men it is seventeen grams a day (recommended thirty to thirty-eight grams a day). This difference may seem like a lot, but seven grams of fibre roughly translates into a seventy-gram serving of whole wheat pasta (approximately two-thirds of a cup) along with a fruit and a serving of tomatoes. As we will see in later chapters, fibre is a terrific food for your microbiota and aids your whole body, not just your brain.

In addition to affecting CVD and resulting strokes, the microbiota may also play a role in post-stroke recovery. When the microbiota is measured in post-stroke patients, it shows signs of dysbiosis within twenty-four hours of the stroke, and the worse the stroke, the larger the change in microbiota composition. This is an interesting finding, but we have no idea what the implications are. However, there are some more concrete clues that the microbiota may affect the outcome of the stroke. We know that the microbiota has a large impact on the immune system and resulting inflammation (see Chapter 11). Tissue damage following a stroke frequently occurs due to inflammatory damage by the immune system. Here, antibiotics can play a positive role, as stroke victims are often given antibiotics to control post-stroke infections. In animal studies, treatment with antibiotics reduced the brain injury, due to the decrease in the activation of key immune pathways. This included an increase in cells that dampen inflammation (called regulatory T cells), and a decrease in those that cause more tissue damage (delta gamma T cells). Antibiotics can't be used to prevent strokes in people for many reasons, including general antibiotic resistance and the profound upsetting impact of antibiotics on the microbiota. However, other tools are becoming available to identity the microbes involved in stroke and, accordingly, manipulate the microbiome to curtail strokes and/or improve recovery.

JET LAG AND SNOWBIRDS

A dream retirement often includes visions of long periods of luxurious, exotic travel. Without the burden of work or dependent family members, you may finally have the time to explore some of the many wonders and corners of the world. In Canada, a major national pastime is called "the flight of the snowbirds," referring to the many retirees who flee their snowbound home cities to spend the winter on a warm beach, returning like clockwork only when spring has sprung. However, long-distance travel comes with a major discomfort, jet lag. Groggy head, lack of energy, severe sleep deprivation, random junk food cravings, and increased appetite . . . yuck, you know the feeling! For the first few days, it is hard to really enjoy the beach; you may even fall asleep mid-day and get a severe sunburn to painfully remember the trip by. Jet lag and other disruptions of the body's internal clock, or circadian rhythm, are not just associated with these short-term changes, but also with higher rates of obesity, diabetes, and cancer. As we will see later, these diseases all have a microbial link. What has become apparent in the last couple of years is that the microbiota is tightly linked to jet lag and other body-clock rhythm disruptions in ways we never previously dreamed of.

Circadian rhythm is under the rule of the hypothalamus, an area of the brain that is regulated by light and other factors to control sleep, appetite, and similar daily rhythms. However, we now know that other parts of the body also have their own molecular clocks that sync up to the brain clock. For example, there is a clock that causes the liver and intestine to "wake up" when it is time to digest a meal during the day, or slow down while we are asleep. In 2014, scientists discovered that the intestinal microbiota underwent circadian rhythms of their own. The scientists engineered genetic mutations in some of the molecular clock genes of mice, which impaired the rhythmic oscillations of their microbiota. They found that the body's clocks actually control the microbiota, with cyclical changes occurring in about 15 percent of the microbial species. The scientists could even recognize characteristic "time of day" microbial species. These changes could be overridden by regular feeding hours, meaning that the mice could get back to normal daily body rhythms with regular mealtimes—just as we humans get over jet lag after a trip. If irregular feeding times were used, the main brain clock of the mice became uncoupled from the more local clocks, such as the intestinal clock, which causes the intestine to "wake up" to digest a meal and slow down when asleep. This study affirms what we already suspected, that mucking up the biological clock affects many biological processes, including digestion and the gut microbiota.

Any such changes in metabolic activities—how we break down molecules to get energy—can also lead to obesity and diabetes. For example, scientists transferred feces from a jet-lagged mouse—or from humans (graduate students flying around the world to attend a scientific meeting)—into germ-free mice. If you're curious how a mouse gets jet lag, you don't actually fly the mouse around the world—just shift the lights in the mouse-house by eight hours for three days, then shift them back again. Researchers found that the metabolic symptoms of jet lag (think carb and junk food cravings) associated with obesity and diabetes could be transferred to the newly colonized mice. The infusion of jet-lagged feces disrupted the mice's microbiota enough to affect their metabolism and make them more prone to diabetes and obesity. The scientists also found that feeding mice high-fat diets was detrimental to their circadian rhythms (and microbiota and expanding waistlines). Other mouse studies showed that female mice were more susceptible than male mice to jet-lag changes in the microbiota; that alcohol worsened the changes in both sexes; and that even intestinal infections were increased given disruptions to normal sleep and rest time.

So how can we use this information to ensure we enjoy our next transatlantic holiday to its fullest? Unfortunately, we are still at an early stage of understanding, and since most of the studies have been done in animals we can't draw human connections. However, some themes are becoming apparent. High-fat diets and alcohol both worsen jet lag. Both eating at your regular at-home meal times and exposure to daylight help you and your microbes adjust more quickly to your current beach time. The hope is that as we understand these complex interactions more, we can develop a probiotic or a prebiotic to lessen the sleepless nights that follow travel.

The Future

Of all the body sites discussed in this book, the brain stands to gain the greatest future potential improvement in lifelong well-being and healthy aging thanks to microbiome findings. We've seen that Alzheimer's, Parkinson's, dementia, anxiety, stress, depression, and strokes all have microbial aspects. Through more studies, we'll gain a clearer idea of how well the MIND diet works and be able to design even better diets for our brains. With advances in testing, we'll hopefully be able to accurately profile gut microbiota to identify individuals at risk for cognitive diseases, and in some cases even make an early diagnosis before the damage is done. And with the next generation of probiotics—scientifically engineered "cocktails" of beneficial microbes—we anticipate benefiting from "mind-altering" microbes, thereby gaining more control over—or even preventing—depression and other mental health issues. There may be drugs that target microbial enzymes in order to prevent strokes and cardiovascular disease, and probiotics that can be used to reconstitute good microbes following antibiotics and to help with recovery from strokes. All in all, we can expect to leverage our microbes to fight cognitive decline and hopefully diminish some of the greatest fears associated with aging.

QUICK TIPS

- **Give your brain a workout:** The brain is no different from the rest of your body—it needs exercise to stay healthy. Aerobic exercise gets your heart pumping, which boosts blood flow to the brain. This improves brain function and protects your brain from Alzheimer's and dementia. Exercise also decreases stress and improves mood (and affects the microbes).
- **Trade your T-bone for a turnip:** A diet high in fibre is associated with significantly lower risk of stroke, while red meat increases risk. High-fibre foods good for the brain and intestinal microbes include whole grains, fruit (e.g., berries, tropical fruits, tree fruits), vegetables (e.g., leafy greens, root vegetables), beans, nuts, and seeds.
- **Reset your clock:** Sleep is essential for brain health—not to mention its impact on the microbiota. Lack of regular, solid pillow time (such as when you're jet lagged) can cause problems with memory, thinking, and mood.
- **Enjoy your morning cup of java:** Besides helping with constipation and keeping you regular, coffee also decreases risk for Parkinson's and Alzheimer's.

4

Healthy Smile, Healthy You: The Oral Microbiome

When Antonie van Leeuwenhoek invented the microscope in 1683, he made the seminal observation that there were more "animalcules" (i.e., bacteria) in his mouth than the number of people living in his home country of the Netherlands. It was humanity's first glimpse of microbes, and the stunning realization that our mouths are inhabited by living organisms. As with so many innovative scientific discoveries, at first nobody believed the "crazy Dutchman," as he was called. It took decades of letters to the Royal Society in London describing microbes in great detail to convince the world. As others peered through van Leeuwenhoek's microscopes, his amazing findings were confirmed, and the field of microbiology was born.

The oral cavity—everything in your mouth from lips to teeth, gums, hard palate, and tongue—normally teems with millions of microbes. You swallow billions of bacteria in your saliva every day. The environment of the mouth flourishes with bacteria, given that we constantly provide fluids, nutrients, and a nice warm temperature. Fighting for their own survival just as we do, resident microbes cling to the tongue, inside of the cheeks, teeth, and gums in order to resist being flushed away when we swallow. Relative to the few hundred years of microbiology, this fact is a new realization. We spoke to Dr. Richard Ellen, a retired professor from the faculty of dentistry at the University of Toronto. Ellen is a recognized authority on oral microbial ecology. He told us: "No one ever used the term 'microbiome' back in the 1970s when I started my career." After earning his dental degree, he became a research fellow at Harvard University and Forsyth Dental Center, where he met an inspiring microbial ecologist named Dr. Ronald Gibbons.

Gibbons was interested in how bacteria adhere to surfaces in the mouth and whether similar principles governed the distribution of bacteria that cause disease elsewhere in the body. He sent Ellen off to the library to learn more about the potential for bacterial adhesion in strep throat. Thus began a fruitful collaboration that caught the attention of medical microbiologists. Ellen later applied his expertise to study microbial ecology in periodontology, the specialty of dentistry that studies the supporting structures of teeth such as the gums, as well as diseases and conditions that affect them. In the lab next door to Gibbons, another microbiologist, Dr. Sigmund Socransky, was using the most advanced methods at the time to document the bacterial populations that colonized different oral surfaces. The process involved "picking individual bacterial colonies that grew on nutrient media, then using laborious biochemical tests to identify only those that could grow under lab conditions," Dr. Ellen explained. "This helped us understand a good number of the microorganisms that lived on the tongue, cheek, teeth, and in saliva, but it grossly underestimated their diversity because most of the resident bacteria were not cultivated by these methods." The field has advanced remarkably since then. Dr. Ellen reflected: "I feel ancient, since I started research on these topics when the confident identification of a core but limited cluster of the bacteria in oral samples was just emerging. Today's rapid DNA sequencing and analysis of the oral microbiome seems a bit like interstellar travel to me!"

Scientists and dentists now know that hundreds of oral microbial species generally live interspersed among many different neighbours rather than as a clonal population (think of a city bustling with multitudes of diverse people and buildings). These complex oral communities are fairly stable during an adult's life, but some major changes can occur in older people. Diminished oral microbial communities in elderly patients give pathogens an invitation to colonize and thrive. This affects not only our teeth and gums, but our entire body.

MYTH: Brushing teeth is important only for oral health.

FACT: Brushing teeth promotes health throughout your entire body, including protecting against heart disease and even dementia.

The mouth reflects overall health at any stage of life and represents the critical first contact point between our alimentary canal (the gastrointestinal tract along which food passes from mouth to anus), the immune system, and the outside world. If we envision the gastrointestinal tract as a river, the mouth is the source: the headwaters from which everything flows downstream. In fact, the majority of all systemic diseases (those involving many organs or the whole body) produce oral signs and symptoms.

There is a surprising correlation between the body's microbiota and dementia, as related to oral hygiene. A large twin study showed that early tooth loss (before thirty-five years of age) correlates with higher levels of dementia. The most startling statistic is that individuals who do not brush their teeth daily have a 22 to 65 percent greater risk of developing dementia than those who brush their teeth three times a day!

How could fastidiously brushing one's teeth possibly affect dementia? As one ages, saliva production slows, which enhances low-grade inflammation in the mouth, as saliva is antibacterial; it also reduces one's ability to wash away and swallow pesky oral microbes. Presumably this allows increased numbers of microbes and their inflammatory products to seep into the body's circulation, triggering additional inflammation. Studies show that when antibodies to oral microbes are circulating in increased numbers—an indirect indication, by the way, that these microbes are entering into the body and seen by the immune system—this is directly related to an increased risk for Alzheimer's disease.

Ultimately, oral health is much more important than just an attractive smile. Your mouth is a window into the condition of your entire body and can serve as a critical vantage point to detect and defend against health problems.

Brush, Floss, Repeat

The oral microbiota lives in miniature microbial cities called biofilms. The "skyscrapers" of these cities are built from molecules called extracellular polysaccharides (EPS), which form sticky structures in and around the microbes and protect them from their harsh environment. Biofilms can be pretty much anywhere, but the most common site in the body is the mouth, in the form of dental plaque. Which microbes inhabit the biofilm is important, and it depends on who adheres first. The early colonizers shape which microbes colonize later, i.e., which "neighbours" get to move in, and this sets the composition of the oral "neighbourhood," for better or worse. The most common first colonizers, *Streptococcus* species, help other beneficial microbes subsequently move in. As others colonize and the neighbourhood expands, around one hundred microbial species are found intertwined in the plaque. Microbes usually layer three hundred to five hundred cells deep—a microbial skyscraper indeed!

We've all experienced, to some degree, what plaque buildup feels like: the white or dark-coloured flakes that dental hygienists scrape off our teeth are calculus (tartar)—old biofilm material from bacteria that has calcified into adherent masses. Unfortunately, plaque is constantly building from the moment you finish having your teeth cleaned, and it accumulates in hard-to-clean spots in the six months between visits. Meanwhile, as plaque builds up on and under our gums, direct contact between biofilm and gum tissue causes irritation. The gums inflame (our body's 911 response), and this inflammation summons macrophages and neutrophils: cells designed to damage and kill microbes. However, in many people, prolonged and repeated bouts of inflammation can cause collateral damage to the underlying tissue, and later periodontitis—a destructive form of gum disease. The microbes shift towards more pathogenic ones as the disease advances, which further damages the soft tissues and underlying bone structure.

Surprisingly, the two most common infectious diseases worldwide are not among the usual suspects, devastating pandemics like AIDS, malaria, or tuberculosis; rather they are periodontitis (gum disease) and dental caries (cavities). Nearly all humans have had at least one cavity, and up to half of us experience some form of gum disease. However, unlike the above big three infectious diseases, oral diseases are caused by a mixture of microbes rather than one particular pathogen. People with periodontitis have distinctly different microbes. *Porphyromonas gingivalis* and *Treponema denticola* are often among the microbial culprits behind this disease. Amazingly, germ-free lab animals do not suffer from periodontitis, which makes sense, given that they do not have any microbes.

Older adults are more susceptible to gum disease: The CDC reports that 47.2 percent of adults aged thirty years and older have some form of periodontal disease, as well as 70.1 percent of those over the age of sixty-five. However, it is not a "normal" part of aging: good hygiene and a healthy oral microbiota can prevent this disease. You know the drill: thorough brushing, interdental cleaning (a.k.a. flossing), and regular trips to the dentist. Looking to the future, you may also be able to use probiotics to help "good" microbes colonize the gums and keep pathogens away. Lactobacilli and bifidobacteria are the two genera most commonly found in these probiotics. A normal part of the oral microbiota, they are generally regarded as safe, and can be consumed through oral probiotics and food products. Dairy sources of probiotics include yogurt, kefir, cultured cottage cheese, and buttermilk; and non-dairy sources include fermented vegetables such as sauerkraut and kombucha tea. Use caution, however, as lactobacilli and bifidobacteria can be harmful to people who frequently consume fermentable carbohydrates. These include sugary foods (e.g., cookies, cakes, soft drinks, candy) and less obvious foods such as bread, crackers, bananas, and breakfast cereals. Lactobacilli and bifidobacteria use the sugars from these foods to produce acids. In technical terms, they are incredibly acidogenic and aciduric members (tolerant of a highly acidic

MYTH: Eating sugar causes cavities.

FACT: Cavities are not *directly* caused by sugar, but instead by certain microbes living off the sugar. The microbes produce acid which dissolves tooth enamel and creates cavities.

environment) of the deep plaque that causes dental caries. This means that, through acid production, they can make plaque worse.

Oral probiotics are not commonly used at present in the dental profession, due to lack of compelling clinical trials. They may eventually represent a novel therapeutic strategy to help promote oral and systemic health. Localized treatments with a culture of *Lactobacillus acidophilus* have significantly helped those suffering from periodontal diseases (e.g., gingivitis and periodontitis) to recover. Studies show that the probiotic *Lactobacillus* strains *L. reuteri, L. brevis* (CD2), *L. casei Shirota*, and *L. salivarius* WB21, as well as *Bacillus subtilis,* can improve gingival health and reduce the number of periodontal pathogens. Probiotic gum (see box on page 70, "Chewing Your Way to Dental Health") also represents an easy method to possibly chew your way to improved gum health and decreased periodontal disease. It not only provides an oral probiotic, but the act of gum chewing would also benefit many older people who suffer from reduced salivary flow rates.

As with periodontitis, microbes are at the heart of dental cavities. When the biofilm builds up on the enamel of our teeth, bacteria associated with caries (such as *Streptococcus mutans*) colonize and shift tooth microbiota. The microbial diversity decreases, and a few pesky microbes like *S. mutans* and various lactobacilli increase significantly in numbers (*S. mutans* may multiply from about 2 percent of the microbes to over 30 percent). These microbes hang out on tooth surfaces happily munching on the sugars and other carbohydrates that we feed them every day. In return, they produce acids (lactic, formic, acetic, and propionic), which the microbes can withstand but our teeth cannot. Cavities form when the acids dissolve our enamel and decay the underlying dentin layer. This leads to all-too-familiar symptoms of toothache, sensitivity, and mild-to-sharp pain when biting or consuming something sweet, hot, or cold.

We too often overlook the role of microbes in treating teeth caries, and instead blame the sweets we all love. Germ-free animals on a high-sugar diet do not even get cavities! In addition to well-known dental and dietary habits (i.e., brush, floss, cut back on sugar), an ecologically balanced and diverse oral microbiome might help diminish cavities. Several studies show that consuming probiotics containing lactobacilli or bifidobacteria can reduce the number of pesky mutans streptococci in saliva. *Streptococcus salivarius* M18, commonly found in oral care probiotics, destroys harmful *S. mutans* bacteria and helps neutralize acidity in the mouth. Boosting microbial defences can combat the root cause of cavities.

We all know that regular brushing is important to oral health. However, it seems that not all toothpastes are equal: those that affect the microbiome population may enhance health benefits. In a recent double blind randomized study, 111 healthy adults brushed their teeth twice a day with fluoridated toothpaste for four weeks, after which those participants were divided into two groups: one that continued to use the regular fluoridated toothpaste, while the other began using Zendium, a fluoridated toothpaste that also contains three enzymes that generate antimicrobial products as well as three antibacterial proteins. After using either toothpaste twice daily for fourteen weeks, the participants' oral microbes were analyzed. The results were extremely encouraging: the supplemented toothpaste promoted a positive microbial community shift towards more beneficial microbes (twelve microbial taxa, including *Neisseria*), and a decrease in those associated with periodontal disease (ten taxa, including *Treponema*). The toothpaste seemed to boost the mouth's natural defences by relying on natural proteins and enzymes that are similar to those found in saliva. Zendium is available in most of Europe and the Middle East, and you can buy it from Amazon and other online retailers in North America.

Jessica tried it out and noticed that it was a bit different from traditional toothpastes because it does not contain the foaming agent sodium lauryl sulphate. Although there wasn't that clean-feeling froth, it tasted good and left her mouth feeling clean. As scientific advances in molecular techniques boost our ability to understand the oral microbiome at the species level, we can expect to see even more specific microbe-friendly toothpastes in the future. Given that oral health largely depends on the balance between health-promoting and disease-associated bacteria in the mouth, we can increasingly tip the scale in our favor by applying supplemented oral products.

CHEWING YOUR WAY TO DENTAL HEALTH

Given that periodontitis, caries (cavities), bad breath, and other oral issues are all caused by troublesome microbes taking up residence in the mouth, can we rid ourselves of these problems by displacing the bad bacteria with "good" microbes? Oral probiotics may be able to help with this. One microbe now being used as a probiotic is *Streptococcus salivarius* K12. This beneficial microbe produces bacteriocins (toxins) that kill other microbes associated with both periodontitis and halitosis (bad breath). It was declared safe for people in a randomized clinical trial where ten billion microbes were ingested every day for twenty-eight days. More expansive clinical trials are underway to prove its efficacy, but in small trials it reduced bad breath and decreased throat infections (e.g., strep throat and tonsillitis).

In a clever twist, *S. salivarius* is marketed as a probiotic gum called BLIS K12. You can order this chewing gum in two flavours (spearmint-peppermint or raspberry-pomegranate). It has no sugars or artificial sweeteners and contains five hundred million probiotic bacteria per piece. Jessica was curious about the product and ordered a spearmint-peppermint pack online. At first, she tried one piece per day and chewed it for five to ten minutes after brushing her teeth before bed (as suggested). The gum tasted good, though it lost its minty taste rather quickly. The main snag is that at $13.50 for a pack of eight pieces, it was rather pricey to purchase on her grad student budget.

Given its expense, Jessica would consider chewing it on a case-by-case basis, such as on the way home from dental appointments to help recolonize beneficial oral microbes on her freshly cleaned teeth. It might also be helpful when taking antibiotics—which are intended to attack the body's microbes. For older adults, the chewing gum represents a potentially easy way to deliver probiotics throughout the entire mouth while boosting saliva flow.

Other types of probiotic chewing gum available contain *L. reuteri* ATCC 55730 and ATCC PTA 5289, and may decrease levels of pro-inflammatory cytokines, therefore contributing to better oral and systemic health.

Spit and Polish

Saliva plays a vital role in maintaining oral health as it lubricates your mouth, helps you taste and digest your food, and contains defensive compounds that help control excessive buildup of microbial plaque. It's also teeming with helpful microbes: roughly five hundred million bacteria per teaspoon. Good saliva flow thus delivers moisture and nutrients to your microbes, and buffers as well as flushes away acid produced by cavity-causing microbes on the teeth. We each have a unique microbiome within our saliva that remains fairly stable over time. But as we age, our saliva can dry up: hyposalivation and xerostomia (dry mouth) are present in one-third of older adults and in 40 percent of those over the age of eighty. A dry mouth makes chewing, eating, swallowing, talking, and even whistling—clearly the most essential activity on this list!—more difficult. Without saliva's defensive properties and nutrient delivery, the good and protective microbes can't thrive, which makes us more vulnerable to pathogenic infections.

There is no significant decline in salivary flow for healthy, unmedicated older adults, so dry mouth is not an inevitable nuisance in later life. To find relief if you suffer from it, it is recommended you sip water throughout the day to keep the mouth moist and during a meal to make food taste better. Avoid heavily caffeinated and sugary drinks that can dry out the mouth. The National Institutes of Health recommends sugarless gum or hard candy to boost saliva flow; they even specify that citrus, cinnamon, and mint-flavoured candies are "good choices." There are also over-the-counter saliva substitutes, moisturizing sprays, and gels that can help temporarily moisturize and lubricate oral tissues. Perhaps a probiotic gum could help boost oral microbes and stimulate saliva: Studies show that *Lactobacillus* species *L. reuteri* (DSM 17938 and ATCC PTA 5289), *L. rhamnosus* GG (ATCC 53103), and *L. rhamnosus* LC705, as well as *Propionibacterium freudenreichii* ssp shermanii JS all have promise in diminishing the risk of hyposalivation and the feeling of xerostomia (dry mouth) by replenishing one's supply of microbe-rich saliva.

Many patients and doctors do not know that over four hundred frequently prescribed common medications cause dry mouth. Risk increases when taking four or more medications per day. If this is an issue for you or a family member, talk to your doctor about adjusting dosages and medications to catch the culprit.

Microbe-Breath

Can't you just use mouthwash to kill all the bad oral microbes? Antonie van Leeuwenhoek tried the original mouthwash experiment centuries ago. He knew that both alcohol and vinegar could kill the "animalcules" he scraped out of his mouth. So he did what any good scientist would do: Using himself for an experiment, he tried gargling with alcohol and vinegar and then examined the microbes in his mouth. That didn't work: the microbes were unaffected, likely because microbes are encased in a strong shield of protective biofilm casing. His experiment holds true today: Common over-the-counter mouthwashes generally are only partially effective. While they are great for rinsing away food particles missed by brushing and flossing, limiting plaque above the gumline and freshening odour, they do not rid our mouths of cavity-causing microbes or those deep below the gumline.

MYTH: Mouthwash achieves the same effect as brushing.

FACT: Mouthwash is merely an add-on—thorough brushing and flossing do a much better job of preventing plaque than rinsing alone. Some mouthwashes kill both harmful and beneficial microbes.

Dr. Ellen explained that commercial mouthwash was initially marketed to "kill the germs that cause bad breath," which fell under cosmetic concerns. Mouthwashes did not need to be approved by the FDA because the companies did not assert specific "health" claims. As researchers and clinicians began to learn about dental plaque and its relationship with gingivitis, clinical models for mouth rinses began to emerge. The first antimicrobial mouthwash that proved to prevent plaque-induced gingivitis was chlorhexidine, which later became a prescription drug in North America. "It tastes terrible, but it works," Dr. Ellen explained. "It prevents the accumulation of plaque. In those without brushing skills, such as elderly patients who cannot manipulate the instruments needed to clean their teeth, it can help reduce plaque and lower risk for gingivitis. It is not a good idea, however, to be on it continuously. And rinsing with chlorhexidine does not reach below the gums to affect destructive forms of gum disease."

What about over-the-counter antimicrobial mouthwashes? Like chlorhexidine, these indiscriminately kill both good and bad microbes. They also do not necessarily prevent periodontal disease. Dr. Ellen cautioned that consumers need to read the disclaimers in small print on the bottom of the package to learn about the product's limitations. While generally harmless, long-term use may result in some negative side effects including teeth stains and antimicrobial drug resistance.

Keep an eye out for probiotic mouthwashes—they are an emerging and perhaps more effective oral hygiene treatment than traditional models. Published results of preliminary studies show that probiotic mouthwash containing three different oral streptococci reduced plaque and *S. mutans* associated with cavities. There are many options commercially available: Talk to your dentist before changing your oral hygiene routine. Ultimately, Dr. Ellen advised, today's mouthwashes cannot replace manual cleaning: "Mouthwash is topical, temporary, and does not reach critical areas if someone has periodontal pockets, in between teeth, or in fissures." It is just one component of oral hygiene, and secondary to keeping your teeth clean through brushing, interdental cleaning, and routine professional visits.

AN ONION A DAY KEEPS EVERYONE AWAY

We all have mornings when we need to brush our teeth before greeting any fellow humans. If you have a dry mouth while sleeping, this can cause bad "morning breath." Halitosis (chronic bad breath) occurs in about 25 percent of people and significantly affects quality of life. Microbes living in deep crypts, or valleys, of the tongue cause bad breath. These microbes make 150 different volatile molecules—meaning that they evaporate into air—that stink, including sulphur compounds such as hydrogen sulphide (rotten egg odour) and others that smell like rotten cabbage, decomposing seaweed, fish, and garlic. Yuck! No single microbe is responsible for bad breath; it is a mixture of microbes living together on the tongue.

Older adults particularly suffer from bad breath due to hyposalivation (less saliva to wash away these malodorous microbes and their compounds). Furthermore, poorly fitted dentures allow pockets of bacteria to grow in the spaces which, like grooves in the tongue, can lead to microbial production of smelly molecules.

Traditional methods to control bad breath include teeth brushing, dental floss, tongue scraping, mouthwash, and drinking ample water. All of these produce limited success. Recent clinical studies suggest that replacing the smelly bacteria with good ones may be a more effective way of controlling bad breath. Specific probiotic strains that can alter halitosis include *E. coli* Nissle 1917, *S. salivarius* K12, and *Weissella confusa* isolates. Remember that the probiotic gum BLIS K12 contains five hundred million *S. salivarius K12* bacteria, making it an easy method for fresh breath support. *E. coli Nissle 1917* can be found in oral supplements, but those should be used only with extreme caution: One study found that if both the microbiota and immune response are defective, there can be potentially severe adverse effects. *Weissella confusa* has also been suggested as a probiotic, but it can cause sepsis and other serious infections in humans and animals.

These two examples remind us of the need to approach probiotics with caution and a close eye on the scientific literature. As we mentioned in the opening chapter, the FDA regulates probiotics as a food, not as medication. This means that probiotic supplement companies, unlike drug companies, do not have to scientifically demonstrate that their products are safe or effective. The products are not rigorously approved to prevent or treat specific conditions, which leads to abundant misinformation and false claims. Before you begin to use a new probiotic, research all the specific strains in the product. A quick Google search often leads to the National Center for Biotechnology Information, which has rigorous scientific information publicly available. We discuss how to read up on specific medically tested probiotics in detail in Chapter 14.

Smoker's Mouth

Smoking has profound effects on the oral microbiota as they are directly exposed to highly concentrated doses of smoke. In biofilms of the subgingival area (under the gums), smoking favours the early colonization of pathogens that cause periodontal disease. Smoking also prevents normal microbes from colonizing early enough to take up the space of unwanted pathogens inside the mouth. Given their unhealthier oral microbiome and the direct negative effects of tobacco smoke on the tissue itself, smokers' mouths do not heal as well when their owners seek treatment for periodontal disease. An estimated 42 percent of periodontitis in the United States is attributed to smoking, and, despite all the known dangers, there are forty-five million American adult smokers and one billion smokers worldwide.

Oral microbial communities are less stable in smokers. A lack of healthy and diverse microbes contributes to pathogen colonization in the nasopharyngeal area (head and neck). This leads to additional upper respiratory problems (e.g. ear and throat infections), especially in older adults more susceptible to infection, and causes a buildup of bacteria, which travels from the throat to the middle ear. To add insult to injury, additional gum disease in smokers means that they are much more likely to have bad breath due to their pathogenic microbes.

The damage caused by smoking accumulates, meaning that the longer you smoke, the greater the risk for these ailments. Studies show that using tablets containing *L. salivarius* WB21 can improve gingival health in high-risk smokers and reduce the number of periodontal pathogens in plaque. *L. salivarius* is found in live-culture dairy products as well as probiotic supplements. Beneficial microbes in oral probiotics may help strengthen the barrier against invading pathogens that cause periodontal disease and upper respiratory infections. Of course, quitting smoking represents the best defence and brings immediate health benefits at any age—even in later life. For the sake of your microbes *and* personal health, take advantage of counselling and of smoking cessation medications such as nicotine in gums and patches if you smoke.

MYTH: Brushing your teeth immediately after smoking can make up for some of the harmful effects on your teeth and gums.

FACT: Smoking kills beneficial microbes in your mouth that protect against dental disease and bad breath. The destructive microbes settle right back in the mouth immediately after brushing.

Exciting Times Ahead

This is a thrilling time for oral health as we begin to realize the major role of oral microbes. Recolonization of the mouth with healthy microbes to establish strong microbial communities can play a major role in the future in both the dentist's office and personal hygiene. Commercial products such as probiotic gum and supplemented toothpaste are on the rise to boost the oral microbes. We expect that probiotic mouthwashes will help reduce the number of pathogenic bacteria associated with bad breath, dental caries, and periodontitis.

With the awareness that gum health is critical for lifelong vitality, the medical community will exercise increased vigilance toward the microbes involved in gum disease. Since the mouth is much easier to access than other body sites, we are likely to profile older adults' microbes in order to identify specific risks and recommend targeted interventions. We will also be able to expand personalized treatment to alter the microbes and boost helpful bacteria through diet, prebiotics, and probiotics. Now that is something to chew on!

QUICK TIPS

- **Brush, floss, rinse, repeat:** Early tooth loss is linked to higher levels of dementia. Brushing your teeth regularly (ideally three times a day) may decrease your risk for dementia and Alzheimer's.
- **A new generation of probiotics:** Promising oral health products often include lactobacilli and bifidobacteria. You can incorporate these "mouthy microbes" into your diet through dairy products and fermented vegetables (e.g. cultured yogurt, kefir, sauerkraut), or consider

taking a supplement that includes specific strains. Supplemented chewing gums and toothpastes may also help with bad breath and improve oral health. Using these enhanced products on a regular basis, particularly following antibiotic treatment and after any dental visits, may help recolonize healthy microbes for immune support.

- **You know the drill:** Regular tooth brushing, interdental cleaning, and dentist visits will drastically reduce risk of gum disease and tooth decay as you age. Flossing is particularly important to keep plaque-producing microbes in check.
- **Flossed in translation:** Talk to your dentist and doctor if you are experiencing dry mouth. Medications are often the culprit, and other drugs may be substituted or dosages changed. If medications cannot be avoided, it is advised to drink plenty of water, chew sugarless gum (consider a probiotic variety) and avoid tobacco and alcohol. Perhaps try a probiotic (e.g. *L. reuteri* and *L. rhamnosus*) after taking antibiotics.
- **Malodorous microbes:** Alcohol-based mouthwashes cannot penetrate microbes, and antimicrobial mouthwashes kill off both helpful and harmful microbes indiscriminately. This allows the microbes that cause bad breath to flourish without competition. To treat bad breath, floss and brush routinely to keep microbes in check, and perhaps use a probiotic-infused mouthwash or chewing gum containing *S. salivarius*.

5

Take a Deep Breath: The Lung Microbiome

Take a really deep breath. Hold it. Now breathe out. What happened? Even though you had some conscious muscle control of your breathing just now, it's unlikely that you regularly consider the precise mechanics of this autonomic process. It goes like this: As you inhaled, your diaphragm muscle—located just under the lungs, at the bottom of the ribcage—contracted and pushed itself downward. This increased the space in your chest cavity for your lungs to expand and fill with air, a movement aided by the intercostal muscles between your ribs. Your expanded lungs then sucked in air from outside the body through the mouth and nose. Air travelled down your windpipe and into your lungs, passing through your bronchial tubes into the alveoli (tiny air sacs). You also sucked in microbes from the surrounding air. The oxygen then travelled through the thin walls of the alveoli into the surrounding capillaries (blood vessels) to diffuse throughout your entire body. At the same time, carbon dioxide moved from the capillaries into the alveoli. Exhaling relaxed your diaphragm and intercostal muscles, which reduced the space of your chest cavity and forced carbon dioxide and moisture out of your lungs and windpipe.

This entire process happens effortlessly somewhere between seventeen thousand and thirty thousand times per day—that's twelve to twenty times per minute on average, when at rest. Without it, we would die. While we can live without food for many days and water for a few days, we can only live without air for a few scant minutes. Every day we breathe in about eight thousand litres of air. The maximum amount of air your lungs can hold at once is about six litres (three large soda bottles).

We tend to think of the lungs as an inverted Y structure with two lobes at the end of the central windpipe. If we look very closely, however, we see that it is an intricate structure with thousands of tiny alveoli branches to increase the surface area of our lungs to exchange oxygen and carbon dioxide. If we flattened out all these tiny airways, the surface area would be thirty times that of our skin!

Aging Lungs

What happens to our lungs as we age? First of all, maximum lung function (the maximal force you can generate when breathing in or out) naturally decreases with age. After the age of thirty, the rate of airflow through the airways slowly declines, as well as the amount of oxygen diffusing from the alveoli into the blood. Later on, other structural changes start to play a role in lung capacity. The ribcage can become thinner and change shape due to shrinking muscles, thereby altering the bones' ability to respond to the expansion and contraction of the lungs and diaphragm with breathing. With nowhere to expand into, the diaphragm weakens, which in turn decreases the capacity to inhale and exhale to our fullest. The muscles and tissues keeping the airways open can lose elasticity, and the alveoli can lose their shape and become baggy. The part of the brain controlling breathing may no longer send as strong a signal to the lungs in old age (due to age-related decline), and nerves in airways that trigger coughing may become less sensitive to foreign particles. The immune system can also weaken,

MYTH: All older people lose lung capacity, which is why they can become unable to carry out daily activities.

FACT: Healthy young adults have extra lung capacity, which begins to decline around the age of thirty. But as we age, that surplus enables the healthy lung function needed for daily tasks to continue.

meaning that the body is less able to fight off lung infections (e.g., pneumonia and bronchitis) and other diseases. As a result of these changes, older adults are more at risk for shortness of breath, low oxygen levels, and abnormal breathing patterns such as sleep apnea.

Despite these declines, older people should be able to maintain enough lung function for everyday activities. In our youth, we have "extra" lung function: this is why healthy, younger individuals can tolerate the surgical removal of an entire lung and still breathe reasonably well. In healthy individuals, even these age-related changes seldom lead to serious decline. In fact, poor heart health and obesity are more often the culprits for breathing problems than changes to the lungs and their surroundings. Exercise is thus essential to both overall fitness and breathing. Studies have shown that exercise and aerobic training can improve lung capacity, even in very old populations. Aerobic workouts can improve lung capacity by 5 to 15 percent as exercise "works out" two of your most important organs: the lungs and heart. Regular activity can improve endurance and reduce breathlessness by increasing oxygen capacity. Not smoking is the number one way to minimize the effect of aging on the lungs. More than any other demographic, older adults need to be aware of the need to stand up, move around, and breathe deeply and regularly—especially during illness or after surgery. Lying in bed or sitting for long periods of time at any age allows mucus to build up in the lungs, putting the lungs more at risk for infection.

MYTH: The lungs are sterile and devoid of microbes.

FACT: Our lungs are exposed to upwards of a million bacteria each day and most likely have their very own, essential, microbiome.

Exploring the Lung Microbiome

Every time you breathe in, more than just vital oxygen is entering your body. The air around us teems with millions of microbes, mixing with some of the microbes in your mouth that ultimately make their way through your windpipe and downwards. We have a strong defence system against unwanted microbes, including the lungs' surface, which is coated in mucus, a slimy substance designed to trap microbes. Cells of the lung push that mucus upwards into the mouth, where you swallow and digest it. This is called mucocillary action. Think of the last time you had a cold and coughed up a lot of slimy mucus—that was your body's way of expelling unwanted bacteria and cellular debris. In addition, there are special cells called alveolar macrophages that patrol the lower airways of our lungs looking for harmful microbes to engulf and destroy (more on the immune system in Chapter 11).

But breathing involves much more than just the lungs. There are two main parts of the respiratory tract: Your nose, pharynx, and larynx constitute the upper respiratory tract; and the trachea, bronchi, and lungs compose the lower respiratory tract. All our classic science textbooks say that the lower airway is sterile, and thus no home to microbes. To learn about the history of this paradigm we interviewed Dr. James Hogg, emeritus professor of pathology at the University of British Columbia and long-time lung researcher at St. Paul's Hospital in Vancouver. "That sort of thinking is all gone," he confirmed. "The lung is not sterile! That light went on recently—certainly after interest in the microbiome came along." When asked why the sterile lung idea was so prevalent, Dr. Hogg explained that researchers in part contributed to this misconception. It is part of a well-established, widely used technique to wash the airway during a normal bronchoscopy—a test that allows doctors and scientists to examine the airway and lungs. Clinicians use bronchial washing to spray saline solution over the surface, particularly areas that can't be seen, and then collect cells washed off of the surface to examine under a microscope. The problem is that these washings are subject to frequent contamination from the oral cavity as the tube is pushed down the airway, resulting in what everyone always thought was contamination from the mouth.

Scientists also couldn't grow microbes from the lungs to study through traditional laboratory methods such as agar plates—Petri dishes that contains agar, which acts as the solid growth medium—along with nutrients to culture and grow microorganisms. But with the advent of culture-independent techniques—i.e., DNA sequencing—at last we were able to detect microbial signatures deep inside the lungs. As with many discoveries, this claim did not arrive without controversy. Some scientists continued to assert that the microbial signatures were contaminants from the upper respiratory tract due to washing techniques. Dr. Hogg stated that oral contamination remains a very contentious argument and complication. To avoid such contamination, he prefers to freeze the lungs from cadavers before studying them. Better sampling techniques and differing compositions from other oral sites over the past few years have dispelled most of the controversy; however, the process is far from perfect. Dr. Hogg jokingly teased that pathologists have become veritable tissue hounds, constantly scouring hospitals and clinical trials for diverse lung samples in an attempt to enhance our understanding of the respiratory system.

Appreciation for the lung microbiome is starting to spread among clinicians and researchers, but much more hard science needs to be done. "A lot of people are talking the talk, but not walking the walk," according to Dr. Hogg. We are still not able to culture these microbes, so there remains some hesitation to fully accept the "lung microbiome" concept. The existence of microbial DNA doesn't prove the existence of live microbes. The predicted numbers of microbes are incredibly small in comparison to those of other body sites: one thousand times fewer than in the mouth, and one million to one billion times fewer than in the gut. If you follow the math, this works out to approximately two thousand bacteria per square centimetre in the lungs (which may sound like a lot, but is very little in microbial terms). So, what we can *definitively* say is that there are *very probably* microbes in small numbers in the lungs! This claim, vague as it may seem, nonetheless represents a major re-think of the lung microbiome, with significant implications for medical research and personal health.

THE GUT–LUNG AXIS

Much like the gut–brain axis, there also appears to be a gut-lung axis largely mediated through the immune system. Research in Brett's laboratory has shown that certain gut microbes encountered early in life can have a major impact on a person's chance of developing asthma later in life. This has been confirmed in several other studies, and it seems that the early life microbiome affects how the immune system develops, pushing it towards one that is more or less susceptible to later asthma. In addition to these early life gut microbes affecting the lungs via the immune system, there are new reports of oral probiotics, including certain lactobacilli, suggesting that the microbes you ingest can modulate lung inflammation via the gut–lung axis. For example: *Lactobacillus reuteri* and *Lactobacillus plantarum* can decrease allergic airway inflammation, while *Lactobacillus casei* can have positive effects in treating bacterial and viral pneumonia in mice.

Our gradual discovery of bacteria in the lungs means that we are beginning to appreciate their role in healthy lung function. We are starting to see the route microbes take to settle in the lungs: The upper and lower respiratory tracts are similar, so it's possible that many microbes of the lower lungs are seeded by micro-particles drawn deep into the lungs from the upper tract (when we inhaled them), as well as through inhalation and reflux. These lower microbes consist of about 140 distinct families with fairly little microbial diversity, including varieties of aerobes (microbes that tolerate oxygen, as one would expect in the lungs); aerotolerant microbes, which do not utilize oxygen but can protect themselves from reactive oxygen molecules; and anaerobes—because they can't grow in oxygen, they are buried deep in the tissues and other protected areas devoid of oxygen. Among the few things we know lower microbes do is prevent pathogens we breathe in from colonizing by producing antibacterial products that can kill off potential lung pathogens. They also help to shape the immune response of the airway mucosa, thereby helping to control infections and other health risks we encounter throughout daily life.

What exactly are these microbes protecting us from? Given that lungs are constantly exposed to particles from the air, the lung microbiota respond to everything we inhale. Pollutants cause oxidative stress and inflammation in the lungs, which would alter the lung microbiome. Microbiota seem responsive to pollutants. For example, they may break down inhaled pollutants, or modulate how the lung immune system responds to them. People exposed to smoke from both forest fires and tobacco have higher rates of pneumonia caused by *Streptococcus pneumoniae* (*S. pneumoniae*, also called the pneumococcus). Besides the usual safety reasons, this is an additional reason to stay indoors when smoke alerts are released. Buildings filter many of the smaller smoke particulates out of the air, to keep us and our microbes healthier and breathing easier.

Antibiotic medications we ingest also alter the lung microbiome, as do anti-inflammatories and steroids. Given our still-limited knowledge of the lung microbiome, it is too early to say whether or not these changes are detrimental. However, it's safe to say that any upset to normal microbial composition anywhere in the body is associated with detrimental effects—at some point or another. It's likely just a matter of time before we know more about why avoiding antibiotics could improve our lung function, among other things. We also know that lung microbiota are altered in lung diseases such as Chronic Obstructive Pulmonary Disease (COPD), asthma, and cystic fibrosis. While we don't know whether these changes cause the disease or are a result of changing lung environment during disease, we are using this information to find new ways to promote lung health as we age and advise on which activities might exacerbate current conditions and therefore should be avoided.

Smoking

A collection of studies shows that smoking causes differences in the oral and upper airway microbiomes, but does not cause changes in the lower airway microbes. As we saw in the last chapter, smokers have fewer normal, helpful oral microbes to begin with, and more pathogenic ones. This may be related to smokers' higher rates of pneumococcal pneumonia and susceptibility to respiratory infections. The hypothesis is that smoking disrupts the normal microbiota so that it is not as protective against harmful pathogens. Smoking also dampens the immune system and destroys our body's defences in the respiratory tract. This likely leads to increased infection. The good news for smokers is that if you stop smoking, the microbiota revert to their healthier composition.

More puzzling to scientists are the changes that occur in the more distant gut microbiota due to smoking. Smokers have a reduced ratio of *Firmicutes* to bacteroidetes in their gut (which is usually not healthy—see Chapter 7 for more on these key gut bacteria), and reduced numbers of anti-inflammatory bifidobacteria. Smokers also have more pathogenic bacteria in their guts. We do not yet understand exactly why smoking affects intestinal microbes. Scientists hypothesize that it may be because changes in the environment and host (the person affected), impact the immune system and thereby influence microbes in the gut. That smoking is a known risk factor for Crohn's disease speaks to this microbial connection. However, smoking is perplexingly protective against another inflammatory intestinal disease, ulcerative colitis, and for some reason, Parkinson's disease. We have no idea why—but, again, none of this evidence supports taking up smoking as an exercise in preventive medicine.

Chronic Obstructive Pulmonary Disease

COPD refers to a group of chronic lung diseases, including emphysema and chronic bronchitis, that make it difficult to breathe. Though heart disease, cancer, and stroke often receive more attention in the media and medical literature, COPD is a widespread and deadly killer. In 2013,

329 million people, or 5 percent of the world's population, had COPD. It causes 2.9 million deaths per year, with about 90 percent occurring in developing countries. These numbers are staggering: HIV, tuberculosis, and malaria *combined* kill a similar number of people annually. In the United States alone, COPD is the third leading cause of death. It affects more than eleven million Americans, and the American Lung Association estimates that more than twenty-four million of the nation's population have the disease without even knowing it. It can cause serious long-term disability, even death, and currently there is no cure.

Causes of COPD include exposure to air pollution, secondhand smoke, and dust; exposure to fumes from forest fires and polluting open cooking fires; and exposure to workplace chemicals. Smokers are especially at risk for COPD, with about 20 percent of smokers getting the disease, and 85 to 90 percent of COPD patients in developed countries being smokers. When a cigarette burns, it creates more than seven thousand chemicals, and the inhaled toxins weaken and damage the lungs in the same way that other external pollutants can. Smoking has a linear relationship with COPD: the longer you smoke, the greater your risk.

COPD is often not diagnosed until very advanced stages because people do not know the early warning signs. Shortness of breath progressing to a chronic cough and heavy mucus production are symptoms. Frequent respiratory infections, blueness of the lips or fingernails (cyanosis), fatigue, and wheezing are other early warning signs. It is important not to wait for symptoms to become severe, because early detection of COPD is essential to successful treatment. You can't reverse the damage to lung tissue, but you can treat the symptoms and manage the disease. There are several treatment options, including medications, surgery, and other therapies that can improve quality of life. Symptoms wax and wane in cycles, with most damage to the lungs caused during flare-ups. Respiratory viral infections can cause these exacerbations, making the flu shot and the pneumococcal vaccine extra important for COPD patients. Use of bronchodilators and inhaled corticosteroids during flare-ups can help reduce inflammation, decrease mucus production, and make it easier to breathe. The best thing anyone can do to prevent and manage the disease is quit smoking—or ideally never start.

Many studies document major changes in the lung microbiota of COPD patients, especially during COPD exacerbations. As these cycles worsen, the lung defences weaken and dysbiotic changes in the microbiota increase while bacterial diversity decreases. This disturbed microbial imbalance is harmful in general and can make lung inflammation even worse. Scientists have shown that changes in the lung microbiota occur early in COPD, including increases in Proteobacteria and known respiratory pathogens. This hints, but does not prove, that the microbiota are involved in this type of disease. Another clue is that bacterial pathogens, such as those that cause pneumonia, are isolated at a higher frequency in COPD patients. An emerging hypothesis is that the host immune response to the lung microbiota may be part of COPD pathogenesis. Some studies show that treatment with antibiotics for seven to ten days during exacerbations decreases COPD. However, this remains controversial, because little effect was seen in mild or moderate disease states. We may be able to use this information to help identify COPD in its early stages. In fact, one study found that researchers could identify COPD by looking for ten microbial strains in the lungs. Given that an estimated twenty-four million Americans unknowingly have the disease, this method of identification could become a useful tool for early diagnosis and help rule out other lung diseases.

MYTH: COPD is a man's disease.

FACT: While it's historically true that more older men—who as a group generally smoked at higher rates and over a longer time than women—suffered from COPD, deaths resulting from COPD are currently higher in women than men. The reason is threefold: 1) More women have started smoking in the last fifty years; 2) Women are more vulnerable due to smaller lung size and higher estrogen levels (which play a role in worsening lung disease); and 3) Women are often misdiagnosed because many doctors still do not expect to see this "man's disease" in female patients.

Since probiotics can decrease inflammation, could they help with COPD? Maybe. Test tube studies indicate that *Lactobacillus rhamnosus* and *Bifidobacterium breve* can reduce inflammatory responses in isolated macrophages—a type of white blood cell that engulfs and digests harmful particles, especially at sites of infection. *Lactobacillus casei* can improve the function of specialized natural killer cells isolated from smokers. In addition, some probiotics decrease lung inflammation in mouse models of COPD. Keep an eye out for future studies that link the lung microbiota to the progression, exacerbation, and treatment of COPD in humans.

The Angel of Death

Pneumonia is a common lung infection caused by bacteria, a virus, or fungi—most commonly the bacterium *S. pneumoniae*, a normal inhabitant of the oral microbiota that, under the right conditions (such as weakened lungs) can cause this disease. If invading germs get through the upper respiratory tract's defences, they colonize the lower respiratory tract to cause significant damage and inflammation in the lower lungs. This leads to fluid and debris accumulation, and pneumonia in one or both lungs. The infection and its symptoms vary from mild to severe. Chances of faster recovery are greatest in the young, if the infection is caught early and hasn't spread, and in those with healthy immune systems if they are not suffering from other illnesses. Most healthy people recover in one to three weeks, but it can be life-threatening, particularly for infants, young children, older adults, and those with chronic health problems.

MYTH: Everybody who is a carrier of the bacterium *S. pneumoniae* (a common bacterial cause of pneumonia) will get sick from it.

FACT: It is quite common for people, especially children, to carry the bacteria in their throats without getting pneumonia.

Pneumonia causes more deaths than any other infection, killing 3.5 million people yearly worldwide. It is disproportionally hard on older people, whose mortality rates exceed 44 percent for those over the age of eighty-five, in comparison to 5 to 10 percent in all populations. Historically it was called "The Angel of Death" because it can ease the dying process for already weakened and fragile bodies. Lung defences are especially vulnerable if weakened by infection, a waning immune system—which occurs as one ages—and/or smoking.

Pneumonia affects lungs in two ways: lobar pneumonia, which may be present in only one part (lobe) of the lung; or bronchial pneumonia, which may be widespread with patches throughout both lungs. It causes a cough (often with mucus), fever, shaking, chills, and shortness of breath. These initial symptoms can be confused with a common cold or the flu. However, unlike a cold—which left untreated will generally run its course in one to three weeks as the body's immune system restores health—pneumonia often requires treatment for full recovery. Most cases can be eased by drinking plenty of fluids to help loosen secretions and bring up phlegm, which is one reason why fluids are generally recommended for pneumonia and other respiratory illnesses. Also useful are extended rest, over-the-counter medications to control the fever, and if the condition becomes serious, prescription antibiotics. Do not take cough medicine without first talking to your doctor, as coughing is one method the body uses to get rid of the excess mucus and debris and might be helping you get better. In the past, penicillin worked extremely well in treating bacterial pneumonia, but the rise of antibiotic resistance now often requires new antibiotics. In severe cases, people are hospitalized to receive fluids, oxygen therapy, and stronger antibiotics.

You can reduce your risk of pneumonia by taking several simple steps:

1. Get your annual flu shot (see the box on page 92, "Should I Get a Flu Vaccine?"), as the flu makes you more susceptible to pneumonia due to lung damage. The CDC recommends different types of vaccines depending on age (i.e., for children under two and adults over the age of sixty-five) and for those with certain medical conditions who are at higher risk of bacterial pneumonia (pneumococcal). Talk to your healthcare provider to determine which vaccine is best for you.
2. Make it a habit to wash your hands frequently during cold and flu season (viral airborne infections are often spread via coughing, sneezing, and even breathing, which we can pick up from contaminated surfaces and transfer with our hands).
3. Avoid smoking, as we know that tobacco damages the lungs' ability to fight off infection.

External bacteria are behind pneumonia, but do the microbes within our lungs also affect the disease? Recent studies suggest that they may have a larger role than we previously thought. The current hypothesis is that an imbalance in the oral microbiota (dysbiosis) predisposes us to an overgrowth of pathogens, especially the pneumococcus. This leads to subsequent diseases in the lower respiratory tract. Studies have shown that the upper respiratory tract microbiota is very different in those with pneumonia compared to those who don't have the disease. Researchers in one study were able to predict pneumonia by analyzing five specific microbial species found in the upper respiratory

MYTH: Pneumonia only happens in cold-weather seasons.

FACT: Though we've all heard the warning "put on your coat or you'll catch a cold," air temperature in fact has little impact on the ability of microbes to cause infection. The origin of this myth likely stems from the fact that we gather indoors together more frequently during winter months, so microbes can spread more effectively from person to person.

tract—with 95 percent specificity and 84 percent sensitivity. This suggests that the oral microbiota play a major role in lung protection and, ultimately, in seeding lower lungs with pathogens.

We are still at an early stage in assessing how we can use this knowledge to protect people against pneumonia; however, scientists are getting pretty skilled at saving mice. When treated with broad spectrum antibiotics in one study, mice were more susceptible to pneumonia caused by *S. pneumoniae* (probably because their oral microbiota were disrupted). But when scientists then did a fecal microbiota transplant from healthy mice, this effect was reversed! The infected mice recovered from pneumonia. The reasons why this worked are still unclear—perhaps because they recolonized the gut microbiota, or maybe even the oral microbiota (since they did the fecal transfer orally). Either way, this is an exciting finding because it suggests that certain probiotic-like microbial strains might be used in the future to prevent and treat pneumonia.

SHOULD I GET A FLU VACCINE?

Yes!

The influenza virus causes the flu, a viral infection that affects mainly the nose, throat, bronchi, and occasionally lungs. Most of us have suffered from the flu at some point: fever, aching muscles, headache, sore throat, cough, chills, and fatigue. Older adults are particularly susceptible to the flu: 80 to 90 percent of flu-related deaths are in people over the age of sixty-five. This is because the immune system typically weakens with age (making older adults more susceptible to infection), and the flu can worsen other health problems such as heart disease, lung disease, and asthma. The flu weakens respiratory defences, which often leads to secondary bacterial pneumonia—a disease that can be life threatening. As such, in addition to the traditional flu vaccine, there is a federally approved high-dose vaccine for those over the age of sixty-five. It contains four times the regular dosage and provides a stronger immune response and antibody protection.

We generally get our flu vaccine in the fall to prepare for winter, when people spend more time indoors, making it peak infection season. The influenza virus transmits as an aerosol directly between people—more easily accomplished in close quarters than outdoors. Contrary to some anti-vaccine views, you cannot get the flu from the vaccine. It is made with killed—i.e., inactivated—virus; the worst it can do is produce a slight reaction of redness, soreness, and swelling. Recent data suggests that it's the microbiota that may play a role in our body's response to flu vaccine. Microbial products stimulate the host immune response, which then increases the vaccine response, at least in mice. So wear it as a badge of honour if you get a slight reaction to the vaccine! It may signify that your immune system really recognized the vaccine, which could lead to better protection.

You may have heard that in some years the flu vaccine does not work as efficiently as other years. This is because the influenza virus is a chameleon: it constantly changes its surface to camouflage itself from our immune system and try to outsmart our defences. But we formulate vaccines specifically to protect against these ever-changing components, primarily called H and N antigens (think of H1N1). Every February, the WHO looks to the southern hemisphere to see which flu strains are prevalent there, and makes their best estimate of which strains will appear in the fall in the northern hemisphere. This leaves six frantic months for vaccine companies to ramp up production of that year's vaccine. Some years they get it just right and the vaccine provides great protection; other years, the virus changes during those six months and different strains arise, thereby making the vaccine less effective. This process is closely watched by the WHO in order to begin the process all over again to formulate flu vaccines in time for the southern hemisphere's winter flu season. In either case of effective or less effective flu vaccine years, it is recommended to get yearly vaccines for residual protection from the primary strain. Vaccines even provide cross-protection from other strains that arise that year or in later years. Even if a vaccine is not fully protective, one usually has a milder form of flu.

Still not convinced? Consider this: There are hints (in mice) that the vaccine may help stave off dementia and other inflammatory diseases (such as cardiovascular disease and type 2 diabetes) by decreasing inflammation damage associated with respiratory infections. When young mice are vaccinated, they are healthier and have fewer chronic diseases as they age. This suggests that one should be vaccinated early and often—yet two-thirds of Canadians and close to 60 percent of American adults do not get the flu vaccine. Everyone over the age of six months should get their annual vaccination; the rare exception being those with severe, life-threatening allergies to the vaccine.

Looking forward

Recognition of the lung microbiome and its relationship to the oral microbiota opens up a whole new way of looking at lung diseases. New generations of probiotics may soon be developed to reduce risk for respiratory conditions such as pneumonia. The other central player in lung diseases, inflammation, also affects the microbiota. Encouraging early studies suggest that some *Lactobacillus* probiotics have an impact on reducing lung inflammation. We may also be able to diagnose diseases such as COPD and pneumonia in the future by analyzing a person's microbiome and looking for key species. This means less of a need for chest x-rays, alongside faster interventions resulting from our ability to identify respiratory diseases at an earlier stage than we were previously able to.

QUICK TIPS

- **Know the warning signs:** Early symptoms of pneumonia and COPD can often seem like a regular cold or age-related changes. Seek immediate medical attention if symptoms persist. Early detection is essential to effective treatment.
- **Take a breath of fresh air:** Minimize exposure to smoke and harmful pollutants as much as possible. If you smoke, the best way to improve your respiratory health and boost your lung microbiota is to quit.
- **Consider probiotics:** Taking lactobacilli products can help manage allergic lung disease. It may even help with COPD symptoms to reduce inflammation.
- **Shots all around:** Get your annual flu shot. Due to increased risk of bacterial pneumonia over the age of sixty-five, also get the one-dose pneumococcal vaccine (if applicable) to protect against illnesses such as meningitis and pneumonia.

6

Belly Bugs: The Stomach Microbiome

Eating is a fairly simple process. Food goes in your mouth, you chew it into smaller pieces, and swallow. But have you ever thought about what happens next: How you actually digest your food so that what you eat benefits your body? The stomach can hold roughly one litre of food. It is extremely acidic (a pH level of 2, the same as battery acid) due to the hydrochloric acid needed to further break down food into its components. Digestion in the stomach is aided by proteins called proteases and other digestive enzymes that chop up larger molecules such as proteins and complex carbohydrates. But digestion doesn't stop there: Following this acid bath, food is pushed farther down into the small intestine, where most of the action actually occurs. Enzymes disassemble fats, proteins, and carbohydrates into their components so that the micronutrient building blocks the body needs can be absorbed and passed into the bloodstream, which sends the nutrients to other areas of the body. In the large intestine, trillions of microbes devour what we can't—mainly fibre and other prebiotic carbohydrates. The intestinal microbiota synthesize a variety of vitamins for us, including vitamin B12 and vitamin K.

MYTH: All digestion occurs in the stomach.

FACT: Some digestion occurs in the stomach, but food passes through a series of gastrointestinal (GI) tract stations on its way through our body: the mouth, esophagus, stomach, small intestine, and large intestine.

Because of its extremely low pH, we long assumed that no microbes could live in the stomach—that it was a sterile environment similar to the lungs (another view that is changing, as we just saw in the previous chapter). It had been thought that to protect the body from infections, this high acidity purposefully killed most microbes that were swallowed during eating. However, this line of thought changed in 1982 with the groundbreaking discovery of the spiral-shaped bacterium *Helicobacter pylori* (*H. pylori*), which lives in the stomach. As we shall see, *H. pylori* has rewritten the book on gastric health, disease, and treatments.

Through recent DNA sequencing technologies, we are beginning to realize that there is indeed a small stomach microbiome entirely distinct from the gut's. There are anywhere from one thousand to ten thousand microbes per millilitre in the stomach. This may sound like a lot, but remember: In the gut, or lower intestine, there are billions of microbes per millilitre. These stomach microbes are acquired from ingested food, the mouth, lungs (swallowing mucus), and from the upper intestinal tract (which is directly connected to the stomach). The most prominent microbes detected in the stomach are the usual suspects: Proteobacteria (including *H. pylori*), *Firmicutes*, bacteroidetes, *Actinobacteria*, and *Fusobacteria*. Like most microbial populations in the body, they vary extensively from person to person. For bodies colonized with *H. pylori* (which is over half the world's), this organism is the dominant bacterium in the stomach. It also influences the other microbes by altering stomach inflammation, which then changes the environment of the stomach and the resident microbiota. While we can say with certainty that there are microbes in the stomach, their exact functions are still being explored. And as our knowledge of this microbiome grows, so will our ability to use it to improve health and medicine.

Ulcers and Microbes

When Brett was growing up, his father had ulcers (a sore that develops on the lining of the esophagus, stomach, or small intestine) that were supposedly caused by too much worry and stress. Everyone nodded in agreement with this diagnosis and instructed him to find ways to relax and avoid spicy food. At the time, the only way to treat the symptoms of ulcers was to take antacids such as Pepto-Bismol. Brett still remembers the big pink bottles filled with chalky liquid that his father drank daily to alleviate symptoms and presumably decrease the damage being done to his stomach. The active ingredient, bismuth subsalicylate, coats irritated tissues in the esophagus and stomach lining to help protect them from stomach acid. While we still use these products today to diminish symptoms of an irritated stomach, what we now know about *H. pylori* has changed everything.

MYTH: Spicy foods cause ulcers.

FACT: Hot sauce lovers, rejoice! We now know that the microbe *H. pylori* causes most ulcers, along with the use of anti-inflammatory pain medications such as aspirin and ibuprofen. Spicy foods may aggravate symptoms in some people, but they won't literally eat a hole through your stomach. Ulcers are caused by inflammation in the stomach lining: When stomach acid seeps through the protective mucus barrier, a lesion grows and further damages the tissue.

We spoke to Dr. Martin Blaser, a distinguished professor in New York University's Departments of Medicine and Microbiology and a world expert on *H. pylori*. He is a physician by training whose evolving research interests in microbiology over the past forty years have focused on bacteria of the human microbiome, including *Helicobacter* species. Dr. Blaser repeatedly noticed that people with certain diseases had lower rates of *Helicobacter*, which "led to the idea that *Helicobacter* wasn't so bad at all. Its effect is mixed, with both costs and benefits for its human hosts." As a medical doctor, he knew that there was a link between acid reflux and asthma. As a microbiologist, he saw that if *Helicobacter* decreased, asthma increased. By integrating medical practice and microbiology research, he became open to the idea that *H. pylori* was protective against asthma.

Most clinicians operate under the outdated mindset that if a patient with stomach ulcers has *Helicobacter*, you destroy the microbes. Dr. Blaser's findings were a huge shift—one that has taken over twenty years to convince various audiences of its efficacy. But as was the case with the truth about *H. pylori* and ulcers, the tide is changing. As he told us, "Now when I speak to microbiologists about potential protective elements of *H. pylori*, they respond: 'Of course, that makes perfect sense.' However, in the clinical community, doctors are still searching for and destroying *H. pylori*. If a patient's symptoms are at all related to *H. pylori* and gastrointestinal issues, their doctor tries to destroy the microbe. This is a huge extrapolation based on limited data." He likened *H. pylori* to the canary in the coalmine: It's a disappearing organism, and without its protective and effective efforts we may suffer the consequences.

A NOBEL EFFORT

The story of how *H. pylori* was discovered is a classic underdog story that stood science on its head. As previously discussed, the stomach was long thought to be a sterile, even hostile, acidic environment. However, as far back as 1875, forgotten German papers described spiral bacteria seen in the stomach, although no one could grow them. Italians also saw them in 1893, and Polish researchers reported them in 1899. In the 1950s, researchers conducting an extensive American study of stomach biopsies failed to see any of these microbes, further cementing the concept that the stomach was germ-free. Just a few years later, in 1979, Dr. Robin Warren—a gastroenterologist in Australia—observed spiral shaped microbes and convinced a medical fellow working with him, Dr. Barry Marshall, to go after them. They were convinced that these spirals were associated with ulcers and they needed, somehow, to grow these "unculturables" to know more.

In a story remarkably similar to Alexander Fleming's discovery of penicillin, Marshall accidentally left some plates on his laboratory bench top for five days over the long Easter weekend. When he came back, what do you think he found? Microbes: The same he'd seen in every patient who had ulcers and gastritis (stomach inflammation). Marshall and his team were thus convinced that ulcers were not caused by stress and spicy foods, but by this microbe. Although confirmed within months by several microbiologists, their initial paper in 1982 was met with skepticism in the scientific community. So to prove the role of *H. pylori* in ulcers, Marshall made the ultimate sacrifice for science—he performed an experiment on himself.

He first proved that he did not have any *Helicobacter* in his stomach by way of an endoscopy. He then scraped two Petri dishes that contained a growing strain of *H. pylori* isolated from a patient. Wisely, he checked first to make sure that this strain was sensitive to antibiotics so he could treat himself. Marshall assumed that it would take years before he saw anything, but instead, after five days he became sick with nausea and vomiting. Again, using an endoscopy, he found that he had gastritis (possibly an acute, self-limited infection), and was colonized with *H. pylori.* He treated himself with antibiotics and recovered fully. This caught the world's attention. The scientific community gradually accepted his finding that a microbe could cause gastritis—and now all doctors know that gastritis, ulcers, and most gastric cancers are caused by this microbe. Doctors Marshall and Warren were awarded the 2005 Nobel Prize in Physiology and Medicine for their discoveries.

H. pylori colonizes over half the world's population and is considered the most common human pathogen on the planet. Having co-evolved with humans for the last sixty thousand to one hundred thousand years (see the box on page 101, "Ötzi the Iceman"), it is usually acquired early in life, often from our mothers. The microbe is corkscrew-shaped with a spiral and special tail (called a flagellum) that allow it to burrow deep into the mucus of the stomach, where the pH is much more reasonable for a long-term stay (about 6 to 7). *H. pylori* produces urea, a chemical base that neutralizes the stomach acid even further. To top it off, the microbe has a specialized syringe-like system that allows it to pump virulence factors directly into stomach cells to reprogram them to suit the microbe's purpose. Clearly *H. pylori* used those one hundred thousand or so years of coevolution with humans wisely!

ÖTZI THE ICEMAN: FROZEN IN TIME

In 1991 German hikers found a human body encased in ice in the Alps. After it was chipped out of the glacier, the body was taken to an Austrian lab where they determined he had died 5,300 years ago (between 3,359 and 3,105 BCE). He was named Ötzi, after the nearby Öztal Valley. Since his discovery, much has been learned about this individual. Scientists sequenced his genome and found that he is closely related to people living in Corsica and Sardinia today. Ötzi had sixty-one tattoos, making him the oldest tattooed mummy. He had whipworm (an intestinal parasite), was lactose intolerant, had cardiovascular disease, poor knees from extensive walking (perhaps he was a shepherd), and was involved in copper smelting. We also know a fair bit about Ötzi's last day alive: his final meal was red deer and bread, and he was killed by an arrow to the back or a crushed skull—either way he met a violent death. Luckily for us, he was rapidly covered by snow and promptly freeze-dried by the environment, which prevented destruction of his body by predators and decomposition. Ötzi even has the distinction of having his own institute, the Institute for Mummies and the Iceman, located in Bolzano, Italy, where they continue to study him.

After several years, researchers realized that his stomach hadn't disintegrated as assumed, but was hidden within the body. What's more, scientists could get microbial genetic material from the stomach. Using this precious material, they were able to sequence the entire genome of an old *H. pylori* strain! It more resembles an Asian strain than the European strains we see in that area these days. This has provided clues about how *H. pylori* evolved in humans. The current European strain originated from an African strain, which must have arrived after Ötzi met his sudden death over five thousand years ago.

H. pylori gets a bad rap these days since we pay a great deal of attention to the 10 percent of people harbouring it who get peptic ulcers—although ulcers are not an inevitable outcome. When this connection is present, there is the worry that prolonged ulcer formation and genetic mutations can cause cancer. Only 1 to 2 percent of those colonized with the strain get gastric cancer; however, *H. pylori* has the distinction of being the only microbe that is a Class A carcinogen—that puts it in the same cancer-causing category as cigarettes—because *H. pylori* causes 80 percent of gastric cancers (the third most common cancer worldwide). A big risk for stomach cancer is age. The longer *H. pylori* lives in the body, the greater the chance it will cause cancer due to cell turnover and mutations. Most people diagnosed with this cancer are in their late sixties through their eighties.

MYTH: You'll know if you might have stomach cancer because you'll feel pain.

FACT: People with stomach cancer rarely have symptoms during early stages of the disease. This is one of the reasons why early detection is so hard. In the United States, only about one in five stomach cancers is found at an early stage before it has spread to other parts of the body.

So if *H. pylori* causes all these problems, why not just get rid of it, as people were doing for years? Because, as Dr. Blaser cautioned, the organism has so many benefits that the drawbacks may be a small evolutionary price we have to pay. Epidemiological studies suggest that having *H. pylori* protects against asthma, obesity, allergies, various infections, and inflammatory bowel diseases.

This protection has, in many ways, allowed our species to survive; he explained, "Humans and *Helicobacter* are very clearly a story of co-evolution. This co-evolution has late-life costs to people, which are gastric cancer to a major degree, and ulcer disease to a lesser degree. However, a critical question remains as to early-life benefits from *H. pylori*. It likely provides benefits from fighting off infection. We have lots of evidence that a healthy microbiome is important, of which *H. pylori* is an essential part. It's something that has lived in many of our species for at least two hundred thousand years. Now *H. pylori* is present in fewer than 5 percent of children that were measured in the United States, Germany, and Scandinavia. This is a *huge* ecological change in just a few generations. One that's frightening. We are monkeying with our heritage." Dr. Blaser emphasized that we have to better understand who is truly at risk for gastric cancer instead of indiscriminately killing off the organism that might cause it. Those who are at a high risk of gastric cancer should clearly get rid of *H. pylori* if colonized, but those with low risk should consider prioritizing its protective effects. Given its protective boost against asthma, we may increasingly question how to return *H. pylori* to future generations.

The second reason to not get rid of *H. pylori* is its incredible persistence in the body. There is no vaccine to block it—it's resistant to up to seven different antibiotics—and there are no good therapies to completely remove it. Current *H. pylori* eradication therapies use two different antibiotics and a proton pump inhibitor (PPI) to decrease stomach acidity. Tens of millions of people worldwide take PPIs, making them among the most profitable drug classes in the world. The current thought is that PPIs are safe and can't hurt, but this approach may need revising. PPIs quite clearly have a strong impact on the stomach microbiota; just one of their effects is increasing infections with *Clostridium difficile* and pneumonia. Plus, they are over-prescribed and not as effective as we'd hope: Over 70 percent of prescriptions may be inappropriate, and even their hammer-like effect eradicates only 50 to 70 percent of *H. pylori*.

Third, there is emerging evidence that the microbiota may also play a role in gastric cancer rates even separate from *H. pylori* alone. Models using mice show that altered microbiota cause different disease levels when infected with *H. pylori*. In Colombia, people living in the coastal regions have a much lower rate of gastric cancer than those living at high altitudes. They also have different diets: The coastal communities eat more fruits, vegetables, and seafood, while the higher altitude populations eat more potatoes and fava beans. A recent study compared the gastric microbiota of people from two Colombian towns—one at a high altitude in the Andes, the other a coastal village. Twenty participants from each village were matched by age and sex. Those who lived in the high-altitude village had a twenty-five-times higher risk of gastric cancer than those living in the coastal village. Not surprisingly, people from the same village had more microbiota similar to each other than to people from the other village. The researchers also found very significant differences in the microbiota between the two populations. Two microbial groups (*Leptotrichia wadei* and a *Veillonella* species) were significantly more abundant in the high-altitude population, and other microbial groups were much more abundant in the coastal village participants. These differences make sense given that the two villages had dramatically different diets. However, both populations had shockingly similar *H. pylori* levels. This remains puzzling because they had different diets and levels of gastric cancer. Did the diets affect the ability of *H. pylori* to cause cancer? Were there differences in the *H. pylori*? Could other microbes be involved? While this doesn't prove that microbiota other than *H. pylori* are involved in gastric cancer, it provides compelling reasons for researchers in stomach health and disease to further investigate the role of these and other microbes, as well as diets and foods.

KILL OR KEEP

There are four methods to test for an *H. pylori* infection:

1. **BREATH TEST:** You swallow a harmless pill, liquid, or pudding that contains urea (a waste product the body produces as it breaks down protein). If *H. pylori* are present, the bacteria will convert the urea into carbon dioxide. A device can then detect the carbon dioxide when you exhale. This test can identify nearly all people who have *H. pylori*. It can also be used to check if the microbe has been fully eradicated.
2. **STOOL TEST:** A stool test can detect traces of *H. pylori* in the feces. More specifically, the laboratory analyzes your stool for antigens (foreign proteins) associated with *H. pylori* infection. This test can diagnose an infection and confirm that it is cured after treatment.
3. **BLOOD TEST:** Analyzing a blood sample can detect evidence of an active *H. pylori* infection or a former one. However, breath and stool tests are better at detecting active *H. pylori* infections because a blood test can be positive for years, even if the infection is cured.
4. **BIOPSY:** A tissue sample is taken from your stomach lining during an endoscopy. You'll be sedated for this test as the doctor threads a long flexible tube equipped with a tiny camera (endoscope) down your throat and esophagus into the stomach and the top part of the small intestine. The samples are then analyzed for *H. pylori* infection. Generally, this test isn't recommended solely to diagnose *H. pylori* infection because it is more invasive than a breath or stool test. It is often done for other reasons, including diagnosing an ulcer, treating bleeding, or making sure there is no cancer.

Probiotics vs. PPIs

Given the difficulties in safe and effective elimination of *H. pylori*, there are increasing efforts to use probiotics to help rid the body of this organism. Live microbes taken orally could out-compete *H. pylori* for space and dampen inflammation associated with *H. pylori* colonization. The usual probiotic suspects, *Lactobacillus, Bifidobacter*, and *Saccharomyces boulardii* (a yeast), have been tested, not only because they're common but because they can handle the low pH levels found in the stomach. So far, these studies show a partial improvement in gastric symptoms but not a complete eradication of *H. pylori. (Lactobacillus reuteri* was the best at decreasing inflammation.) When these probiotics were used in combination with the standard triple therapy of two antibiotics and a PPI, the probiotics were more effective at eradication than therapy alone. There were also fewer side effects (diarrhea and abdominal pain), which made it easier for patients to stay on the therapy. An interesting study compared a probiotic yogurt and PPI combination with a treatment of two antibiotics and a PPI. They yielded similar results, but the yogurt/PPI group had fewer side effects. Overall, the addition of a probiotic to the triple therapy may make it a more effective and pleasant treatment for ulcers.

There is no doubt that *H. pylori* is central to both stomach health and disease. However, it is likely that other stomach microbes have an impact, by either affecting *H. pylori* colonization or the inflammation associated with it. It is also likely that more and better probiotics will be found to alter *H. pylori* colonization and gastric inflammation, especially given the increase in antibiotic resistance currently seen in *H. pylori*. As science advances, we will gain a better understanding of what the "normal" stomach microbiome composition is, and whether it affects gastric issues. It is likely that microbes will be increasingly used to counter *H. pylori* and gastric issues.

QUICK TIPS

- **Not just any stomach bug:** Check with your doctor if you suffer persistently and/or severely from any stomach symptoms including poor appetite; abdomen discomfort, fullness after eating a small meal, swelling, or fluid build-up; unintentional weight loss; abdominal pain or discomfort; heartburn or indigestion; nausea or vomiting; and anemia. Unfortunately, early-stage stomach cancer rarely causes symptoms, so having these symptoms may be a sign of advanced-stage illness.
- **Balancing act:** If you have stomach issues and are colonized with *H. pylori*, have a frank discussion with your physician about whether or not to try to eradicate the organism.
- **Pass the probiotics:** If you are on a PPI treatment, consuming a probiotic yogurt could potentially increase the therapy's effectiveness and decrease side effects.

7

Microbe Mecca: The Gut Microbiome

"'Ridiculously healthy' elderly have the same gut microbiome as healthy 30-year-olds." This 2017 headline originating from the University of Western Ontario immediately drew our attention to research headed by Dr. Greg Gloor, principal investigator of a large human microbiota study. The study involved over one thousand "very healthy" Chinese people aged three to over one hundred. What researchers found is that the gut microbiota in the healthy elderly participants were similar to microbes found in healthy individuals decades younger. When kept in good health, it seemed, the gut microbiome could remain robust for over a century. Is this a potential fountain of youth?

The gut microbes of very old people are full of clues in our search for lifelong health and vitality. We can look to the microbiomes of older adults (generally aged 65 and older), including centenarians (over 100 years old) and even semi-supercentenarians (over 105 years old). After the advent of molecular microbiome analysis a decade ago, a series of studies suggested that the diversity of the gut microbiome in older people was diminished. Instead of health-promoting bacteria, scientists observed an increase in Proteobacteria, which are sometimes called "pathobionts" because of their ability to trigger inflammation. As we've seen throughout this book (and we'll discuss in detail in Chapter 11), prolonged low-grade inflammation is strongly associated with the aging process—so this finding made sense. Scientists suggested that increased inflammation is partly caused by increased oxygen production; which, through reactive oxygen that accompanies inflammation, boosts the number of bacteria—called aerobic or facultative anaerobes—that can tolerate such oxygen, including harmful Proteobacteria. As a result, some beneficial microbes that cannot tolerate oxygen (anaerobes) are killed by the oxygen. In our review of many studies conducted since then, we found that older adult populations *do* tend to have rather dysbiotic microbiomes with a decrease in overall diversity. There is a distinct microbial signature among those older than sixty-five that includes marked decreases in beneficial microbes and increases in more harmful bacteria.

In the very old, however, this rule doesn't seem to hold, as diversity *increases* again. Centenarians' microbiomes, like those of younger populations, are dominated by beneficial *Firmicutes* and bacteroidetes (we'll discuss these later on in this chapter). Because semi-supercentenarians are quite rare as a population, it's harder to gather data about them: About 1 in 70,000 people will live to be more than 105 years old. But from the data we've been able to gather, these individuals, too, have an increase in numerous beneficial microbes (*Akkermansia, Bifidobacterium*, and *Christensenellaceae*). We now see that the gut microbiome does not follow linear changes with age: The "downward spiral" we assumed is actually a more complicated trajectory after the seventh decade of life. Since many people now live well beyond sixty-five years, there are ever-more opportunities to see why these changes occur in later life and to understand how we can nourish our microbes well before old age to help promote health and longevity.

Gut 101

While people often think of the gut as only the intestines, the gastrointestinal (GI) tract is in fact a long tube that runs through us from the mouth to the anus. It is divided into various sections including the stomach, small intestine, large intestine, and colon. Although this bodily throughway is internal, the contents of the intestine are not technically inside us. Like the skin, a more obvious barrier between our insides and the external world, the lining of the gut serves as an external surface of the body—it simply runs through the middle rather than coating the outside. Also like the skin, the gut is an important permeable gateway between us and the world, including microbes.

Think of the gut microbiome as an ecosystem: Many different microbes living together harmoniously, most of the time, that can directly and indirectly influence each other. This ecosystem is characterized by a decrease in oxygen the deeper the descent down the intestinal tract (there is virtually no oxygen in the large intestine). Anaerobes are the predominant bacteria, especially in the colon where they make up 99 percent of the bacteria. Microbes that can tolerate some oxygen (called facultative anaerobes) are found in larger numbers higher up the intestine. As in all robust ecosystems, the large diversity of microbial species in the gut is essential.

We can see how different parts of the GI tract have distinct environments, and correspondingly different microbes. As we saw in Chapter 6, the stomach itself (close to the top of the GI tract and very acidic, with a fair bit of oxygen, which is swallowed with food) contains very few microbes (about 10^3 to 10^4 bacteria/gram). Next up (actually, down) is the small intestine, which has less oxygen and is more basic in pH. It is divided into three parts: the duodenum, jejunum, and ileum. The jejunum has about 10^5 to 10^6 microbes and the ileum 10^8 to 10^9 microbes, while the large intestine contains about 10^{11} bacteria/gram of intestinal contents. Finally, we reach the finishing area, where we find the largest numbers of bacteria in—and on—the entire body; about ten trillion in the lower bowel, better known as the large intestine. It is depleted of oxygen and has a basic pH. This remarkable concentration of microbes at the end of our digestive process explains why microbes make up about 30 percent of fecal matter. There are one hundred billion microbes in a single gram of feces—over ten times the entire population of humans on this planet. These microscopic gut microbes collectively weigh as much as two to five pounds, which is roughly the weight of your brain.

Given these staggering numbers of gut microbiota and their many functions, some scientists consider the microbiome the "newest discovered organ."

How do we build up billions of microbes in the intestinal tract? The moment you are born, you receive your first and very best birthday present from your mother: A big dose of vaginal and fecal microbes. Although this isn't something you'd put on your birthday wish list any other time in life, these microbes are an essential part of health in early life as they kickstart your intestinal microbes and all their essential duties. See *Let Them Eat Dirt* by Brett and Marie-Claire Arrieta if you are interested in the role of microbes in early life and how it affects child health and disease. Most new parents know how important a healthy digestive system is to their baby's well-being—and have likely seen first-hand what a "healthy" intestinal microbiome looks like. Remember diaper duty when you introduced solid food into your child's diet? This dietary change causes a major remodeling of the intestinal microbiome.

In the first years of life we're exposed to many kinds of bacteria that can inhabit our gut microbiome. By adulthood, we harbour a multitude of diverse microbes including bacteria, viruses, fungi (yeast), and protozoa. The microbial composition remains fairly constant for most of adult life unless you do something drastic, like become a vegan or move to another country—more on that later. But our gut ecosystem isn't immune to change, and disturbances can have major influences on microbial composition similar to those in early years, when we first ate solid food. As we've seen, antibiotics, for example, can decimate our microbiome like a nuclear bomb because they are not species-specific. They don't just attack the bacteria making us sick but many harmless resident bacteria as well. And like our nuclear analogy, the fallout effects are long-lasting: Recent research shows that the effects of a course of antibiotics on the microbiome can be seen at least a year after taking antibiotics. Some reports suggest that microbial changes up to four years later are apparent.

The Gut's Microbial Village

Why is it important to protect the village of microbes in our gut? For starters, without them we'd have a hard time simply surviving, since gut microbiota play a huge role in breaking down food. The majority of the food we eat simply cannot be broken down without microbes. This is especially true of complex plant sugars—the polysaccharides found in grains, legumes, vegetables, fruits, and other plant foods—which are converted into short-chain fatty acids (SCFA), which serve both as an energy source for intestinal cells and important signalling molecules. More on this later. The beneficial effects of SCFA produced by microbes are many. They help to decrease obesity and insulin resistance, shore up the gut barrier, serve as energy sources for other microbes, reduce inflammation, and send signals to our body promoting the immune system and its development. Diet is a good way to boost your SCFA production. High-fibre diets with low amounts of fat and minimal meat are associated with much higher levels of SCFA and other beneficial microbes. And not surprisingly, they are associated with increased overall health and longevity (see the MIND diet, Chapter 3).

Another important role of gut microbes is competitive exclusion, or the out-competing of pathogens. Research in Brett's and others' labs have shown that normal microbiota can be quite effective at preventing intestinal infections caused by *Salmonella* and pathogenic *E. coli*. In animal models, fecal transfers—teeming with gut microbes of animals resistant to infections—into susceptible animals can enhance their resistance to infections. Some microbiota species even produce molecules that inhibit the virulence mechanisms of *Salmonella* and *E. coli* (i.e., diarrhea), thereby decreasing their ability to cause disease. Perhaps these resident microbes block diarrhea in order to prevent themselves from reaching the light at the end of the tunnel—thereby saving themselves.

Protecting the diversity and density of the microbes themselves is key to gut health, but there's another layer equally important to balanced digestion: the gut's thick mucus lining, which helps keep microbes inside our intestines where they belong. Mucus is a composed of glycoproteins—a combination of complex sugars (carbohydrates) linked to proteins—that coat the intestinal wall. This lubricates it as well as providing a barrier between intestinal cells and the microbiome. By contrast, the mucus layer in the small intestine is thin, perhaps because there aren't as many microbes. The mucus lining has two layers in the large intestine: A very tight one nestled against the intestinal epithelial cells forming a barrier to microbes, and a looser outer mucus layer that is embedded with microbes.

Some of these microbes are able to break down mucus and use it as a food source. *Akkermansia muciniphila* is one such microbe and is thought to be beneficial in preventing intestinal diseases that are associated with thinning of the mucus layer. By eating the mucus, this microbe triggers a feedback loop that actually increases mucus production and thickens the mucus layer, which promotes overall gut health. When the mucus barrier is thinner, intestinal permeability increases. More microbes are able to penetrate the gut barrier, resulting in intestinal inflammation and dysbiosis that can also trigger increased tissue damage and, potentially, disease. As we will cover later in this chapter, many gastrointestinal diseases are associated with thinner mucus layers.

Your Body's Built-In Detox

In our journey through the body's microbiomes, we've already seen how intricately connected all these discrete bacterial communities are. The gut microbiome is no different, and because of its robust size of one hundred trillion bacteria, it influences—and can even improve—health conditions in sites quite distant from it.

Adjacent to the GI tract itself are two important organs that do their own kind of digestion, the liver and kidneys, which are heavily impacted by gut bacteria. The liver filters blood coming from the digestive tract before passing it to the rest of the body, metabolizes drugs, detoxifies chemicals, plays a role in fat metabolism, and produces proteins important to blood clotting. Healthy kidneys filter about half a cup of blood every minute, thereby removing wastes and extra fluids to produce urine. If either of these hardworking organs cannot function properly, it can wreak havoc on our bodies.

Liver diseases are a broad category of conditions that damage the liver and prevent it from functioning well. The most common conditions include cirrhosis (chronic liver damage from a variety of causes, including alcohol consumption, that lead to scarring and liver failure); Hepatitis A, B, and C (serious viral infections that attack the liver and lead to inflammation); hemochromatosis (an inherited condition caused by excessive iron absorption); and non-alcoholic fatty liver disease (the accumulation of liver fat in people who drink little or no alcohol). All types of liver diseases are associated with microbial dysbiosis, increased intestinal permeability—which, as discussed earlier, leads to an increase in the leakage of harmful inflammatory bacterial products into the body—and accompanying chronic inflammation.

DRINK TO YOUR HEALTH—THE SKINNY ON RED WINE

Alcoholic liver disease (which encompasses fatty liver, alcoholic hepatitis, and chronic hepatitis with liver fibrosis or cirrhosis) is caused by chronic excessive consumption of alcohol. The risk for disease increases the longer one drinks and the more alcohol one consumes. While drinking alcohol in excess is unhealthy, moderate consumption of red wine has been associated with significant health benefits, including protection from cardiovascular disease and type 2 diabetes. In Chapter 3 we saw that red wine is a component of the Mediterranean and MIND diets, which lessen the risk of neurological diseases such as dementia.

The major beneficial components in red wine are thought to be polyphenols, which are strong antioxidants that protect the body from reactive oxygen damage. They are found in the skin and seeds of grapes incorporated into the wine. White wines have fewer polyphenols than red wines: With one glass of white wine (120 millilitres, approximately 4 ounces) you would ingest about 60 mg of polyphenols, while a (recommended) glass of red wine of the same size would deliver about 210 milligrams.

It turns out polyphenols also have a beneficial effect on gut microbes, as about 90 to 95 percent of the polyphenols are metabolized in the colon. When patients with metabolic syndrome—a disease associated with obesity and diabetes—consumed one 272-millilitre (roughly 9-ounce) serving of red wine daily for thirty days, they saw large increases in the beneficial microbes bifidobacteria and *Lactobacillus*, which decrease intestinal permeability. There were also increases in the butyrate producing *Faecalibacterium prausnitzii*, and a decrease in the inflammatory Proteobacteria. Several other studies indicate similar beneficial changes in the gut microbiome associated with red wine consumption. In addition, it is thought that the microbial breakdown of polyphenols by gut microbes results in the production of beneficial metabolites, which positively affects human health. One more reason to raise a glass of red to your health!

Non-alcoholic fatty acid liver disease (NAFLD) is a catch-all for the spectrum of liver diseases including steatosis, non-alcoholic steatohepatitis (NASH), hepatocellular carcinoma, fibrosis, and ultimately cirrhosis (scarring of the liver) associated with liver failure. NAFLD is the most common type of chronic liver disease in Western countries, affecting about 30 percent of the adult population. Indeed, there are more than three million cases per year in the United States. The cause of NAFLD is unknown, but risk factors include obesity, gastric bypass surgery, high cholesterol, and type 2 diabetes. No standard treatment exists, so doctors usually try to treat the underlying condition(s) instead. Improvements have been shown from lifestyle changes such as increased exercise, weight loss, lowering cholesterol and triglycerides, controlling diabetes, and avoiding alcohol.

NAFLD patients have increased intestinal permeability and chronic inflammation. Because the liver clears the blood of potentially toxic compounds, it is thought that increased gut permeability exposes the liver to high levels of bacterial antigens such as LPS that trigger extensive chronic inflammation and resulting tissue damage. Both green tea and coffee inhibit NAFLD progression, which may work through the microbiota. NAFLD patients have a dysbiotic microbial composition, an increase in *Firmicutes* and decrease in bacteroidetes.

We may be able to use these marked differences in gut microbiome to better detect NAFLD, which typically is not identified until the disease is well advanced. Even then, diagnosis requires an invasive liver biopsy. Researchers at the University of California San Diego reported that in a small clinical trial the unique microbial makeup of a patient's stool sample (which is typically thought of as a "snapshot" of their gut microbiome) can be used to predict advanced NAFLD with 88 to 94 percent accuracy. The team found that patients with advanced NAFLD tend to have more Proteobacteria and fewer *Firmicutes* in their stool samples than those with early-stage NAFLD. At the species level, one major difference was in the abundance of *E. coli*, a bacterium that can trigger inflammation: These bacteria were three times more common in advanced NAFLD patients than those in the early stage. Determining exactly who is at risk or has the disease is critically needed for prevention and treatment. Stool-based tests may be key to detecting NAFLD simply based on microbial patterns.

As discussed above, kidneys are essential to remove wastes and extra fluid from the body. They also remove acid and maintain the balance of water, salts, and minerals in our blood. Nerves, muscles, and other tissues cannot function normally if the kidneys are not able to maintain this essential balance. Chronic kidney disease (CKD) affects between 5 and 10 percent of the world's population, accounting for up to a million deaths worldwide per year. It is a nonspecific disease characterized by the slow deterioration of the kidneys over months to years. The kidneys filter waste and excess fluid from the blood. As the kidneys deteriorate and even fail, waste builds up in the circulation and tissues. Late stages of the disease, such as renal failure, often require a kidney transplant or dialysis, which requires a machine to filter excess water, solutes, and toxins from the blood—the function healthy kidneys would normally perform.

Like so many of the chronic inflammatory diseases discussed in this book, CKD also appears to have a significant microbial contribution. Its progression is directly related to microbial composition and is correlated with high blood pressure, diabetes, and cardiovascular disease. The microbial effects are also similar to other conditions: Marked dysbiosis in the gut microbiota, including fewer SCFA-producing bacteria (which dampen inflammation), accompanied by an increase in gut permeability. This results in bacterial products such as endotoxin (LPS) entering the bloodstream, increasing the body's inflammatory responses and damaging the kidneys. Several studies suggest that repairing gut dysbiosis may improve CKD, including by following high-fibre diets; the use of some probiotics improved blood markers associated with CKD.

Another common condition is a kidney stone: A small hard formation caused by the deposition of chemicals (calcium phosphate and, mainly, calcium oxalate) in the kidneys. Up to 12 percent of Americans are affected by kidney stones in their lifetimes. Once you've had one (which you will never forget), you're 50 percent more likely to get another within the next ten years. The stone-causing chemicals originate in our diets. It is thought that when microbes such as *Oxalobacter* and *Lactobacillus*, which normally break down oxalate salts, are in lower quantities, stones are more likely to form. A recent small study of twenty-three individuals with kidney stones and six controls found that those with stones had a markedly different gut microbiome. They had higher *Bacteroides* and *Prevotella*. Focusing specifically on *Oxalobacter formigenes*, another study of 247 patients with kidney stones and 259 controls found a strong inverse correlation with kidney stones and this bacterium. In other words: People lacking this microbe had a much higher risk of kidney stones. In another study, three people were fed this bacterium once, along with a high oxalate diet, which resulted in decreased urinary oxalate secretion. Antibiotic usage is also associated with increased risk of kidney stones.

While we are still awaiting establishment of a direct causal relationship between gut microbes and kidney stones, these highly suggestive studies indicate that the gut microbiome is involved. This may present novel future solutions to avoid this extremely painful disease.

All Guts, No Glory

Certain specific digestive tract conditions, such as irritable bowel syndrome (IBS), are immediately and directly affected by the gut microbiome. IBS affects about 10 percent of the population and is most common in women and younger people; however, it is one of the most commonly encountered functional GI disorders in older people. Its symptoms are vague and many—recurrent abdominal pain as well as changes in bowel habits, including bloating—and while treatment can help ease symptoms, it cannot be cured. IBS does not directly affect longevity, but it has a major impact on quality of life and can cause chronic pain for years.

IBS is a complex syndrome. Several lines of evidence indicate that the microbiome is involved, whether through a dysbiotic microbiome and/or decreased microbial diversity (no studies have yet been able to identify a characteristic microbial signal associated with the disease). One reason that researchers believe microbes have a role is because IBS often appears following gastroenteritis (called post-infectious IBS) and, to a lesser extent, antibiotics. Fecal transfers of IBS patient feces into germ-free mice transfers many symptoms of the disease, including diarrhea and anxiety. Like many intestinal diseases, there is an increase in intestinal permeability in patients and accompanying low-grade systemic inflammation. It is thought that this contributes to gut–brain signalling, perhaps by altering neuromuscular responses, gut motility, and pain perception. This may contribute to associated depression and anxiety, though it is not proven.

Although there is no cure for IBS, pilot studies indicate that probiotics that decrease intestinal inflammation show some promise as a treatment to alleviate symptoms. In one study of forty-four patients with IBS, researchers found that *Bifidobacterium longum* NCC3001 reduced the depression often associated with IBS, but not IBS symptoms or anxiety. Other therapeutic approaches to help manage symptoms of IBS include a high-fibre diet, physical exercise, stress management, diarrhea medication, laxatives, probiotics, and other dietary supplements.

A separate, and more severe, group of bowel conditions gaining increasing awareness in public conversations on gut health are inflammatory bowel diseases (IBD). They are often confused with IBS because they share similar names and some of the same symptoms, but these GI conditions are completely distinct. IBD results in the body's own immune system attacking parts of the digestive system. Approximately 1.4 million Americans have been diagnosed with IBD, and 10 to 15 percent were diagnosed at sixty years of age or older. Diagnosis may take longer and be more challenging in older people because of the large number of conditions common to aging that mimic the symptoms of IBD which doctors need to first rule out. Infections or certain medications, such as nonsteroidal anti-inflammatory drugs (NSAIDs) and antibiotics, can also complicate diagnosis in older people.

There are two IBDs in humans: ulcerative colitis and Crohn's disease. Ulcerative colitis causes inflammation

MYTH: Stress and anxiety cause irritable bowel syndrome (IBS).

FACT: Experts don't know exactly why people get IBS. While stress and depression can make symptoms worse, IBS is not solely a psychiatric illness. There is bidirectional communication between the brain and gut, so symptoms may be caused by dysfunctions primarily in the central nervous system, the gut, or a combination of both.

in the digestive tract, while Crohn's disease affects the lining of the digestive tract, leading to increased gut permeability, inflammation, tissue damage, and fibrosis (the thickening and scarring of tissue). As the name implies, both diseases feature an inflamed intestine that results in diarrhea, rectal bleeding, abdominal cramps, pain, and other serious side effects. Given that they are lifelong diseases, many older people have lived with IBD for many years. The number of hospitalizations for those over the age of sixty-five is greater than in younger populations, accounting for approximately 25 percent of all admissions for IBD. As with IBS, unfortunately, there is no cure and the exact cause remains unknown.

Researchers suggest that a combination of four factors leads to IBD: a genetic component, an environmental trigger, an imbalance of intestinal bacteria, and an inappropriate reaction from the immune system. There are over two hundred human genes associated with an increased risk of IBD, including many genes involved in inflammation and the control of microbial infections. However, together they only account for a small amount of the collective disease risk: About 13 percent for Crohn's and 8 percent for ulcerative colitis. This suggests that there are indeed other factors that contribute to the majority of the causes of these diseases.

The microbiome is now considered a major factor in these diseases, largely because of its interplay with the immune system (see Chapter 11 for more on this), which ultimately affects inflammation in the gut. Immune cells normally protect the body from infection, but people with IBD often lack normal immune responses that keep the gut microbes at bay. This allows microbes to penetrate through the gut barrier, which then triggers intestinal inflammation and tissue damage. IBD patients also tend to have a dysbiotic microbiome—more specifically decreased bacteroidetes, *Firmicutes*, and the butyrate-producing *Faecalibacterium prausnitzii*; and increased inflammatory microbes such as Proteobacteria, and microbes that produce the harmful gas hydrogen sulphide, which damages tissue in the bowel.

There are two compelling lines of evidence that the microbiome is actually involved directly in IBD. First, fecal transfers (which swap gut microbes between hosts) can improve the disease symptoms, especially for ulcerative colitis. Second, studies show that if feces are taken from IBD patients and transplanted into mice without the disease, this causes intestinal inflammation and other symptoms associated with IBD. The current view of IBD is that it is not caused by a single microbe, but that it is a combination of certain risk factors, especially involving the immune system, plus enhanced microbial penetration of the gut, which triggers inflammation. Attempts to cure these diseases thus are now focused on dampening inflammation with anti-inflammatories and improving microbial balance with fecal transfers and other dietary and lifestyle methods.

Stool Transplants

The holy grail of the microbiota world is to identify specific beneficial microbes, grow them in culture, and put them into individuals to better treat diseases such as IBS and IBD and perhaps even promote overall health and longevity. While the process sounds simple enough, as we saw above, defining a healthy aging microbiome is still well beyond our abilities. However, there is one medical procedure that has changed the entire microbiome world, stimulating countless companies to think about microbial therapies: That is, transferring feces from one person into another.

The concept of fecal transfers—also called fecal microbiota transplantation (FMT), fecal bacteriotherapy, or duodenal infusion—is not new. As far back as the fourth century, people in China suffering from severe diarrhea were given oral fecal slurries as a treatment (thank goodness we've advanced beyond that method!). In the sixteenth century, stool and water ("yellow soup") was used as a treatment for various gut diseases ranging from diarrhea to constipation. There are numerous similar reports of fecal uses throughout history. Our modern methods go back to 1958, when Dr. Ben Eiseman of Colorado led a group of surgeons in treating four patients with severe intestinal disease (pseudomembrane colitis) with fecal enemas—and remarkably cured their disease. Recently FMTs have garnered much attention in the medical field as well as the popular press, especially regarding infections caused by *Clostridium difficile* (*C. difficile*, colloquially known as "C. diff").

C. difficile is an intestinal pathogen that causes severe diarrhea by producing potent toxins in the gut. However, as pathogens go, it has trouble establishing itself in the gut if there is a healthy normal microbiome present (competitive exclusion in action). These infections are most often seen in hospitalized patients who are taking antibiotics for reasons such as hip transplants. Surgeons routinely give antibiotics such as clindamycin and other fluoroquinolones to decrease the risk of infections post-surgery. When the drugs wipe out the microbial ecosystem, *C. difficile* can more readily colonize people, triggering severe intestinal disease and, ultimately, even death.

The result is now considered the most common healthcare-associated infection, with over a half million people per year infected in the United States alone. It has a 9 percent mortality rate. Because antibiotics created the conditions to cause infection, treatment with antibiotics such as metronidazole and vancomycin are successful in only about one-quarter of cases. Recurrent *C. difficile* infections, classed as two to three previous standard treatments with antibiotics failing, are common and medically challenging to treat.

Several well-designed clinical studies have shown that FMTs routinely cure more than 90 percent of these infections. These groundbreaking studies demonstrate that simply transferring feces from one person to another can cure a deadly disease. FMTs are now recognized as the clinical treatment of choice for recurrent *C. difficile* infections. About thirty to five hundred grams of fecal slurry is given either by a nasogastric tube (a tube running down the throat to the intestine) or by an enema (injection of fluid into the lower bowel by way of the rectum). Donors are rigorously screened for potential pathogens and their history of antibiotic use, diarrhea, etc. Thus far this treatment seems safe and effective, although we still don't know all the short- and long-term consequences. For example: A transplant from an obese donor into a lean recipient cured one patient of *C. difficile*, but the recipient also gained significant weight. More on the "microbial memory" later in this chapter.

If FMTs work for *C. difficile*, why not use them for other diseases such as IBD? There are literally dozens of FMT clinical trials underway for myriad diseases ranging from obesity to IBS (ClinicalTrials.gov lists 198 studies for FMT at the time of writing). Thus far, in addition to recurrent *C. difficile* infections, the other disease showing significant promise for FMTs is the IBD ulcerative colitis (UC). However, the results are quite mixed among studies, with about 25 percent of IBD patients going into remission due to FMT. IBD is not as susceptible to microbiome competition as *C. difficile* because in IBD, the gut is inflamed and microbiota are already dysbiotic. Incoming microbes from the fecal transfer have to displace the resident ones, dampen the inflammation, and reset the whole gut—which is a big ask.

MYTH: You can give yourself a fecal transfer from a healthy friend or family member.

FACT: Despite the simplicity of the procedure—and the numerous DIY YouTube videos featuring brown mixtures in blenders—do NOT attempt this at home. If the gut is ruptured, you can get sepsis and possibly die of complications. *FMTs are to be performed only in regulated medical settings under careful medical observation.*

As we saw earlier, different donors also seem to have different effects, likely based on their microbiome composition. Some trials are experimenting with blending various donors' feces to provide an even more diverse microbiome. Experiments are also underway that give repeated fecal infusions, combined with pre-treatment antibiotics and/or anti-inflammatories to enhance the chances of the incoming microbes being able to establish themselves. There are initial reports coming from several diseases that FMTs hold promise as an intervention. In a preliminary small trial, FMTs were able to increase insulin sensitivity, which is important for decreasing type 2 diabetes and obesity. Unfortunately, patients weren't followed long enough to measure an effect on body weight.

If the idea of an FMT slurry turns your stomach (or your gut), there is hope on the horizon. Work is underway to pack feces into pills (although you would have to swallow a lot of pills to consume a pound of feces). Even more encouraging is the concept of growing a defined population of human microbes in the lab to give directly to patients. This decreases the risk of unknown viruses and other blood-borne diseases. Dr. Emma Allen-Vercoe and her team at the University of Guelph piloted this and successfully treated two *C. difficile* patients with this synthetic microbe mix. She affectionately calls the procedure "rePOOPulating," and we shall discuss it in greater depth in the final chapter. As we further define certain microbes associated with particular benefits, we should be able to move away from fecal material into clinical-grade cultured organisms. Companies such as Vedanta Biosciences, Inc. are working on specific combinations of live microbes (Live Microbial Products, or LBPs) that can be used instead of crude fecal transfers. Stay tuned!

PASSING GAS

Although flatulence (farting) is frowned upon in most societies, it is a natural process in which microbes play a key role. The term flatulence comes from the Latin word *flatus*, which means "a blowing, a breaking wind." Whenever you ingest and swallow food or drink, a small amount of air also enters the gastrointestinal tract. Once inside, it has to go somewhere—usually either expelled as a burp or fart. Drinking carbonated drinks or chewing gum increases this air intake. The average person passes gas ten to twenty times per day, expelling between 500 and 1500 millilitres daily, or roughly two to six cups. About 99 percent of the gases expelled are odourless, and include nitrogen, oxygen, carbon dioxide, hydrogen, and methane.

The real problem with flatulence is its smell, for which we have microbes to thank. In addition to odourless gases, microbes also produce hydrogen sulphide (which smells like rotten eggs—and is the main culprit for smelly farts), methanethiol (rotten vegetable smell), and dimethyl sulphide (cooked broccoli odour). The smelly compounds are produced in the colon by microbes breaking down complex oligosaccharides (think: plant fibre), which the body is unable to metabolize on its own. Thus, eating broccoli, beans, cauliflower, and other plant fibres can increase the odour of our farts. Vegetarians have increased flatulence given their plant-based diets. Regular flatulence, however, is a good sign, as it indicates good consumption of fibre and a healthy, gas-producing microbiome.

Nourish Your Microbes, Nourish Yourself

The diet of the Western world has shifted drastically over the past fifty years—to the detriment of our waistlines and our microbiomes. On average, Americans eat about 23 percent more calories per day in 2010 than in 1970 according to the US Department of Agriculture. Nearly half of these calories come from just two food groups: flours and grains, and fats and oils. Chicken has topped beef as the most-consumed meat, while in the dairy aisle we drink less milk (42 percent less compared to 1970) but eat more cheese (21.9 pounds per year, nearly three times the average consumption in 1970). Americans now eat 29 percent more grains, mostly in the form of breads, pastries, and other baked goods. In 1999 the average American consumed an average of 26.7 teaspoons per day of caloric sweeteners. This number was down to 22.9 teaspoons per day in 2014—although, remember that these figures don't contain noncaloric sweeteners such as aspartame. While most of the sweetener consumed in 1970 was refined sugar, we now eat a lot more corn-derived sweeteners, including cheap and readily available high-fructose corn syrup.

When you alter your diet, such as switching to a high-fibre/low-fat diet, changes in the microbiome can be detected within three days. It has been suggested that changes in diet can account for over half of the variation in the microbiome in a person (57 percent), while genetic variation accounts for only 12 percent. One great example of how the gut microbiome adapts to diet comes from Japan. We know that a specific microbial enzyme (beta-porphyranase) breaks down a component of seaweed (glycans) and is found in marine microbes. It turns out that this enzyme is also found in a certain microbe (*Bacteroides plebeius*) in the gut microbiomes of Japanese people, but not in North American microbiomes. It seems that eating non-sterile seaweed, as many Japanese do, enabled the transfer of the gene encoding the enzyme for an ocean-dwelling microbe to resident members of the Japanese human microbiome and passed this around Japan, thereby enhancing the population's digestion of seaweed.

The case of seaweed is a specific illustration of how food consumption uniquely altered a community's microbiome. By extension there is concern that broad dietary changes across the general population will cause us, as humans, to lose the microbes that we evolved with since we no longer eat the foods that allowed for our evolution. In a series of intriguing experiments, Dr. Justin Sonnenburg from Stanford University showed that we may be doing serious, irreparable damage to our collective human microbiome. First, he fed "humanized" mice (i.e., germ-free mice colonized with human fecal microbes from a Westerner) a diet rich in plant fibre for six weeks. Next, he fed half of the mice a low-fibre (Western) diet for seven weeks, followed by a high-fibre diet for another six weeks. The control group remained on the high-fibre diet the whole time. The mice were bred, and this was repeated for each subsequent generation. What he and his team found was quite disturbing. At first, the changes in the gut microbiome accompanying the diet change were partly reversible in a single generation (one-third of the original species did not fully recover). But then, with each subsequent generation, a low-fibre diet caused a further decrease in microbial diversity. After four generations, no length of time on a high-fibre diet could recover the microbes that break down fibre. Interestingly, however, researchers could restore the high-fibre microbes by a fecal transfer of normal microbes to the depleted animals.

Collectively this suggests we are driving the microbes that digest high amounts of fibre (which then produce SCFA and many other beneficial effects) to extinction as society continues its low-fibre/high-simple-sugar diet. This effect could compound over just a few generations to the point of becoming irreversible, thus ending our relationship with the beneficial microbes that were so integral to our successful evolution. Losing the diversity and abundance of our ancestors' gut flora could negatively impact health and well-being if our species has fewer and less diverse microbes, making us more susceptible to gastrointestinal disorders, other diseases, and dysbiosis. Perhaps the diverse gut microbiomes characteristic of centenarians and semi-supercentenarians will become impossible to achieve in the future, which could detrimentally affect our prospects for longevity. Scientists are now considering the possibility that we biobank feces from our grandparents for the health of future generations. It may also be important to incorporate high levels of fibre into our lifestyles if we want to save essential gut microbes before it's too late.

The Obesity Epidemic

Obesity was once considered a cosmetic issue caused by overeating or lack of self-control. But now, multiple national and international medical and scientific societies, including the WHO, recognize obesity as a chronic progressive disease resulting from multiple environmental and genetic factors. Obesity is common, serious, and very costly worldwide: One report estimated the global cost of obesity at two trillion dollars annually. More than one-third (36.5 percent) of adults in the United States are obese—a massive increase from 13 percent in 1962—and over two-thirds (69 percent) of American adults are overweight. An estimated 2.2 billion adults are overweight worldwide, which includes 700 million adults who are obese. The disease has a

MYTH: Exercise is the key to successful weight loss.

FACT: To lose weight, eating less is far more effective than exercising more, according to scientific studies. And given that our gut microbes are so heavily influenced by diet (and, to some extent, exercise), we may be able to use personalized microbe-attuned diets for more effective weight loss.

greater cost on our longevity and quality of life, as it leads to other life-threatening diseases including heart disease, stroke, and, most especially, type 2 diabetes.

Obesity has surpassed smoking as the top preventable cause of cancer death: Approximately 20 percent of the six hundred thousand annual cancer deaths each year are now associated with obesity. The disease raises mortality risk for adults of all ages, but this relationship is nearly twice as strong for people under the age of fifty. Obese middle-aged people are more than twice as likely as normal-weight persons to die prematurely. Being slightly overweight in later life is in fact associated with lower mortality risk, but obesity raises mortality risk—especially for coronary heart disease. It also negatively affects cognition and is associated with an increased risk for Alzheimer's disease and related dementia.

The prevalence and severity of obesity is a major cause of present-day attention on the microbiome. In the mid-2000s, pioneering work by Jeffrey Gordon's lab at Washington University revealed that the microbiome plays a major role in body weight. The studies showed that germ-free mice consumed many more calories than normal mice yet gained less weight. They also demonstrated that transferring microbiota from obese mice to thin mice through FMTs was linked to significant weight gain, while transferring microbiota from thin mice to obese mice caused weight loss. This profound finding suggests that the microbiome affects not only weight itself but also metabolism rates and calorie production. Our bodies use the SCFA and other compounds that microbes make for energy.

Fast forward a decade: We have learned even more about these crucial connections but, as with all things in science, also realize that it's more complex that we originally thought. We know that the microbiome breaks down much of our food into energy products such as SCFA that feed the body. If you consume an obesity-inducing Western diet—including processed sugars and refined carbs—that is low in fibre and high in calories, your body selects for a less diverse microbiome. This is because most of the processed food is readily absorbed in the small intestine since it is already broken down, thereby starving the fibre-munching large bowel microbes.

Hence why obese individuals have a different microbiome than those with normal weight: There is a higher ratio of *Firmicutes*, which are very efficient at energy extraction from food, to bacteroidetes, which are enhanced by a fibre-rich plant-based diet. It seems like we need the bacteroidetes to balance out the *Firmicutes*. Multiple studies confirm the association of obesity with gut microbiota composition by studying the *Firmicutes*/bacteroidetes (F/B) ratio, although others point out it isn't quite so straightforward. Weight gain and loss is a multifaceted lifestyle process that cannot be reduced to gut microbiota composition alone.

Recalling the link between the microbiome and inflammation, we see how the microbial dysbiosis observed in obese individuals leads to increased intestinal permeability, which allows microbial products to seep into the body and trigger chronic inflammation. This can lead to insulin resistance (type 2 diabetes), cardiovascular disease, cancers, and dementia. Diets with higher fibre and fewer processed products are thought to help prevent this gut permeability, as well as decrease excess energy production from SCFAs, which helps weight control.

Because of the complexity of the microbiome, there are currently no magic combinations of microbes you can take to create the right ratio of microbes and help you lose weight (although there is a lot of work underway in this area). Personalized diets hold significant promise in tailoring food intake to your microbiome, and it is hoped that approaches incorporating the microbiome can be used to help combat obesity.

YO-YO DIETS

We know all too well that losing weight is really hard. How many of us have made a New Year's Resolution to finally shed those ten pounds, achieved some weight loss after lots of effort, only to gain back the weight we lost—if not even more—a few months later? So we try again next year and the cycle repeats.

This pattern of yo-yo dieting is extremely prevalent in our society. About 80 percent of people with long-term obesity who lose weight through dieting gain it back within a year, and even more weight than they previously started with. Six years after participating in *The Biggest Loser*, a dramatic weight loss reality television show where contestants undergo intensive, monitored dieting and exercise regimens, most participants were back where they started. On top of that, their metabolism had slowed so much that they were in fact burning fewer calories per day than before appearing on the show. Our bodies fight hard against weight loss. This may have served as a buffer in our evolutionary past by helping our bodies cope with a feast-and-famine existence.

Renowned microbiologist and gastroenterologist Dr. Eran Elinav, a scientist at the Weizmann Institute in Israel, experienced this first-hand: "One of the lab's projects started because of my screwed-up lifestyle. I was drinking lots of artificial sweetener drinks and dieting. I suffered from recurrent obesity." Partnering with Dr. Eran Segal and other researchers at the Weizmann Institute, he devised a mouse model to mimic yo-yo dieting in humans. They fed mice one of three options: 1) normal food, 2) a high-fat diet, or 3) cycled between these diets about every five weeks. Mice on the yo-yo diet cycles experienced major problems: They incurred massive microbiome changes (researchers identified 773 bacterial genes that were affected) that lasted about one-quarter of the mouse's lifetime; and those mice gained weight much faster than even the mice who remained on high fat diets (or the controlled regular food).

"The pattern of recurrent obesity seems to have an inherent capability of exaggerating from cycle to cycle," explains Dr. Elinav. "This is why, over the years, we gain a little bit and then a little bit more." These results suggest that there is "microbial memory" in the gut that sabotages weight loss efforts (at least in mice) because of its "memory of previous episodes of obesity—one that is a very disturbed, obesogenic configuration." Even though weight fluctuated widely up and down, the mice's microbiome memory configuration was very persistent.

Using fecal transplants, Elinav and Segal showed that, amazingly, these effects could be transferred into germ-free mice. Fecal matter of the cycling-diet mice promoted faster weight gain in the germ-free mice, which also supports the theory of microbial memory. Although the exact mechanisms of this process remain undefined, researchers found decreases in the flavonoids apigenin (found in artichokes, parsley, and celery) and naringenin (found in grapefruits) in the yo-yo diet mice. These flavonoids are associated with high energy consumption, which can help stave off weight gain. By feeding flavonoids to the mice, their energy consumption rates returned to normal.

Although exciting in theory, we still know very little about how this might apply to humans. Dr. Elinav explained: "We studied mouse weight gain and loss over six months, which is a quarter of mice's lifetime in captivity. A quarter of a human's life is much more complicated to study given that humans live much longer than two years." The data is still to come. Dr. Elinav and colleagues are now following a cohort of humans to study their microbiomes through recurrent obesity behaviour over the course of a year.

Our best takeaway is that in order to maintain healthful changes in the microbiome structure, we have to aim for long-term maintenance of the diet changes that produce it, not cycle on and off. Otherwise our microbes rapidly revert to their previous composition. Moving forward, we can work with our microbes to, hopefully, develop more successful, long-term, and healthful diet strategies to maintain a healthy weight.

Diabetes

In 2015, 30.3 million Americans (9.4 percent of the population) had diabetes, which included twelve million adults over the age of sixty-five. Diabetes remains a leading cause of death in the United States, and direct medical costs were $237 billion in 2017. The disease is caused by a lack of response to insulin, which stimulates the body to take up glucose. This results in elevated levels of glucose in the blood (called hyperglycemia), which then causes damage to the body through tissue damage, blindness, and circulatory issues (e.g., diabetic foot ulcers, which can result in lower limb amputation). People who are overweight or obese have added pressure on their bodies' abilities to use insulin to properly control blood sugar levels, which makes diabetes more common among this population. There are three types of diabetes:

- **Type I diabetes:** Also called insulin-dependent diabetes, this autoimmune condition often begins in childhood. Different risk factors, including genetics and some viruses, contribute to type 1 diabetes. The body attacks its own pancreas with antibodies to the point that the damaged pancreas cells cannot make insulin.
- **Gestational diabetes:** This form only occurs during pregnancy, and researchers do not know why some women develop the disease. Risk factors include being older than twenty-five, having prediabetes or a close family member (such as a parent or sibling) with type 2 diabetes, and excess weight. Women with gestational diabetes develop high blood sugar levels, which can be managed with help from doctors, as well as dietary and lifestyle adjustments. After delivery, gestational diabetes usually goes away.
- **Type 2 diabetes:** Most people with diabetes have this condition. Nearly 10 percent of Americans have type 2 diabetes, and another 86 million have prediabetes (their blood glucose levels are abnormally high but not yet high enough to be full-blown diabetes; a fasting blood sugar level of 100 to 125 mg/dL is

considered prediabetes, while 126 mg/dL or higher indicates type 2 diabetes). People with type 2 diabetes make insulin, but their cells do not use it as well as they should (insulin resistance). Being overweight is a primary risk factor for type 2 diabetes because the more fatty tissue you have, the more resistant your cells become to insulin. However, you do not have to be overweight to get the disease. Additional risk factors include fat distribution (if your body stores fat primarily in the abdomen), inactivity, family history of type 2 diabetes, and age. The risk of type 2 diabetes increases as you get older, especially after the age of forty-five. This may be because we are susceptible to weight gain as we age, and older adults are prone to exercise less and thus lose muscle mass (see Chapter 12 for more on the relationship between mobility and age).

Diabetes can accelerate normal aging processes and shorten life expectancy. Among those aged fifty-five to sixty-four, diabetes reduces life expectancy by eight years. Having the disease makes older people more likely to develop or worsen existing health complications including eye diseases, gum disease, sexual dysfunction, and peripheral neuropathy (which increases fall risk given a numbness or pain in the feet). It can be more difficult for older people to manage self-care including blood glucose monitoring, insulin dosing, and meal planning.

Given the close linkages between obesity and type 2 diabetes, the microbial involvement in this disease is very similar to in obesity. The shift towards increased *Firmicutes*, decreased bacteroidetes, and increased gut permeability contributes to the body becoming increasingly unresponsive to insulin (insulin resistance). Just as with obesity, there are studies indicating that modulating the microbiome improves glucose homeostasis and decreases disease effects. Although not clinically proven, there are reports that prebiotics and probiotics such as *Lactobacillus reuteri* GMNL-263 can decrease the inflammatory responses and improve metabolic control and insulin sensitivity in type 2 diabetes patients.

Beneficial diets rich in fibre also improve insulin sensitivity, probably by restoring the helpful microbes and decreasing inflammation. In a small study using fecal transplants, obese people with type 2 diabetes who received feces from healthy donors had improved insulin sensitivity after six weeks, although there was no observed weight loss. Not surprisingly, multiple courses of antibiotics (two to five, taken over years) were associated with an increased risk of type 2 diabetes, although a single course was not.

Metformin is the most commonly used drug to treat type 2 diabetes and works by lowering blood glucose to inhibit liver glucose production. Several studies indicate that the drug works at least partially through the microbiota, which may be why it has to be taken orally (think access to the gut microbiome). In fact, it doesn't work if injected directly into the blood stream. There are two main mechanisms that may be at work in metformin. First, particular members of the microbiome may modify the drug into a more beneficial molecule that the body responds to positively. Second, several studies support a theory that metformin affects the microbiome composition by pushing it towards a more beneficial anti-inflammatory composition, which then helps improve insulin sensitivity. This beneficial microbial effect can even be transmitted by fecal transfer from metformin-treated patients (with their metformin-altered microbiota) into germ-free mice.

ARTIFICIALLY SWEET

Non-caloric artificial sweeteners have been used extensively in diet drinks and foods, and widely touted as safe substitutes in weight loss programs and diabetes management. Six artificial sweeteners are currently approved by the FDA for use in humans: acesulfame potassium (also called acesulfame K), aspartame, saccharin, sucralose, neotame, and advantame. They are used to sweeten food with fewer calories and carbohydrates than sugar—the sweetening power of most low-calorie sweeteners is at least one hundred times more intense. With the exception of aspartame, none of these sweeteners can be broken down by the body, so they pass through our systems without being digested—hence, no added calories.

But are these products really the panacea for excessive sugar consumption? The question is a contentious one with many differing opinions, and studies on them offer mixed results. A recent study called their safety into question due to effects specifically on the microbiome. Eran Elinav and his team in Israel gave mice three sweeteners in their drinking water (saccharin, sucralose, and aspartame) for eleven weeks (a control group received normal sugar). They found that mice fed any of the artificial sweeteners developed marked glucose intolerance, yet mice given normal sugar had normal glucose tolerance. They also found that treating the mice being given the artificial sweeteners with antibiotics on either lean or high-fat diets blocked the increase in glucose intolerance, suggesting that microbes are involved in glucose intolerance.

Moving from mice to humans, they found that in 381 non-diabetic subjects, those who consumed artificial sweeteners had increased metabolic syndrome parameters, including impaired glucose intolerance. When a group of individuals who normally do not use artificial sweeteners consumed saccharin for seven days, there was a marked shift in the gut microbiome of about half of subjects who responded to the sweeteners by seven days. Their microbiomes resembled those found in type 2 diabetics: Increased *Bacteroides* and decreased *Clostridiales*. When feces from these responders were placed into germ-free mice, the recipient mice had significant glucose intolerance compared to those that received the non-responder feces. This one study suggests that the products we use to control glucose tolerance, obesity, type 2 diabetes, and metabolic disease may actually aggravate risk and symptoms due to their microbial effects. Further studies are needed in humans to better understand short- and long-term effects of artificial sweeteners.

The Personalized Microbe Diet

Perhaps the most exciting promise in gut microbiome research is developing personalized diets based on each person's unique microbial composition. This could help us answer the burning questions of why people respond differently to foods, and why certain diets seem to work well with some people yet not others.

One way we've been able to address this question is by comparing the microbiome composition of individuals to the foods they consume to determine how that corresponds with glucose spikes, which contribute to obesity, type 2 diabetes, and metabolic syndromes. In a landmark paper published in 2015, the team led by Eran Elinav and Eran Segal followed eight hundred people for a week, monitoring all foods consumed (close to fifty thousand meals), tracking glucose levels, and characterizing each person's gut microbiome throughout. They found that particular foods caused different levels of glucose spikes in different people (hinting that standard diets may not work for all). Using machine-learning algorithms, they were able to predict which foods might cause glucose spikes in particular individuals based on their microbiome composition. They then confirmed their predictions in a cohort of one hundred people in a double-blind controlled trial.

These results suggest that each of us responds differently to different foods based on our microbiome composition. It is now possible to predict which foods cause glucose spikes in a person and thereby design a personalized diet based to avoid such spikes. Dr. Elinav explained: "We are now engaged in long-term studies to directly compare gold-star diet recommendations with personalized microbiome diets. In diabetics and pre-diabetics, for example, we can hopefully ameliorate, or even prevent, disturbances in blood sugar control. The jury is still out on long-term benefits of this approach, but in the short-term we're already seeing great success."

This work has now been commercialized by DayTwo, which offers microbiome analysis and personalized diet predictions. Initial indications are that it is more effective at controlling weight gain than general (non-microbial) diet strategies and at preventing glucose spikes associated with type 2 diabetes.

Case Study: Inside Brett's Gut

Given that Brett has spent his life studying microbes, he was curious about which microbes are in his body. And thanks to our current technology, he decided to experiment with three well-known gut microbiome analysis companies to find out. (Disclaimer: Brett has no commercial interest in any of these companies but does know the founders of American Gut and DayTwo, given his research in the field.)

American Gut: Run by Rob Knight, a well-known microbiota researcher, this company is set up primarily for data acquisition. By using vast amounts of data provided by clients, they hope to better understand the human microbiome—think of crowdsourcing for microbes. The test costs US$99 (plus US$25 if shipped to Canada), payable to the University of California San Diego, where Dr. Knight's lab is located. Brett ordered a kit to be sent to him in Canada, but after several weeks no kit arrived. After much back and forth with their helpful team, it turned out that FedEx did not like shipping sterile swabs to Canada.

Finally, after another six weeks, a kit arrived. Brett placed a small amount of his feces in a solution in a tube (usually swabbed off toilet paper) and mixed it up (this preserves the microbial DNA during shipping). You can also get skin and oral microbiota analyzed for additional fees. Fortunately, the post-offices in Canada and the US had no problems shipping feces across the border. After two more months (which is a normal wait time for this type of analysis), the results were ready for viewing via a website.

The results are basically a data readout of your gut microbiome. You get a long list of the gut microbes that are in you, and various graphs of how your microbes compare to those of others. This is great if you are a curious microbiologist like Brett, but there are really no actionable items that can be taken from the results. The results even include the disclaimer, "currently, we cannot tell you what it means if you have more or less of certain bacteria than other people. Gut microbiome research is still new, and we have a lot to learn. Your participation in American Gut will allow us to learn more, and we hope to update you with new findings as they emerge." They do provide lists of certain microbes that are more prevalent in you than in others, your most abundant microbes (Brett had 17.7 percent *Faecalibacterium*), and some moderately interesting dot plots comparing your sample to others, including Amerindians and Malawians (non-Western populations that, guess what, have different microbes). Their website also offers some generic information on the analysis and the human microbiome.

If you would like to sponsor microbiome research, then getting a test with American Gut will help those efforts. However, unless you are a serious microbial geek, the information is not very informative other than at a general interest level.

uBiome: This company is more geared toward personal microbiome testing than data collection (the "u" in the name is actually the micron symbol, hence a play on words). Their kits cost $89, plus $19.99 shipping. Because they are a company, the packaging is slicker than American Gut's. The test follows the same process through a swab of fecal matter. Fortunately, there were no issues shipping samples either way across the border. They go to great lengths to make sure the paperwork is correct and included for shipping the sample back. After the normal two-month waiting period, you get an email inviting you to check out the results.

It does have some fun information once you get past the initial disclaimer: "These statements have not been evaluated by the FDA. This product is not intended to diagnose, treat, cure or prevent any disease." It gives you an overview of all your gut microbes, including the *Firmicutes* to bacteroidetes ratio, which we saw earlier is associated with obesity in some studies. Even though Brett has a normal BMI, his ratio was 1.2:1 (remember he also had a high *Firmicutes* level with the first company, American Gut, too). uBiome didn't detect any probiotics, which was good as Brett doesn't take them. He also scored in the eighty-first percentile for diversity, which was a good sign given how many studies indicate diversity is associated with less disease. There are also some fun bioinformatics prediction tools to play with, such as predicted metabolism based on microbial composition. Much like American Gut, it really just provides a snapshot of your particular microbes on the day of the sample.

Although Brett wanted to do the sampling for both companies at the same time, due to shipping issues it wasn't possible. However, it is fun to compare the results, realizing that one's microbiome changes daily. For example, American Gut had Brett's *Faecalibacterium* at 17.7 percent, while uBiome had it at 12 percent. Even more surprising was that American Gut detected *Akkermansia* (the beneficial microbe that can help thicken the mucus layer of your intestine) at 0.23 percent while uBiome had it at a whopping 8.7 percent. In fact, there were very few similarities between the two, which is likely partly a result of testing at different times, although one would expect more similarity. There are perhaps differences between the way the two companies test and analyze the samples, and it is also likely due to what Brett ate and drank in the intervening time.

DayTwo: A truer taste of the future uses of microbiome testing can be found with DayTwo. As described earlier, this company is focused on predicting glucose spikes based on personalized microbiota analysis. They are a fairly new start-up company. They charge three hundred dollars for their services and are only available in the US, so Brett had to ship his fecal samples through the US Postal System on one cross-border trip. Once you sign up, you fill out an extensive diet questionnaire and provide them with a more extensive fecal sample swab than the other two (about a gram or so of feces in a special tube). Again, in about two months you get the results available via the web or through a specialized App for the iPhone.

This test provided a much more extensive analysis and sequencing with detailed analysis of the microbiome content. For example, Brett's results indicated his *Faecalibacterium* was 9.74 percent (lower than the other two companies) and *Akkermansia* was 0.06 percent, closer to the American Gut result. They also provide lots of interesting information and tidbits about each microbe and graphics to explore the levels of microbes pictorially using various sized circles.

Where DayTwo becomes most useful is predicting what foods will affect your glucose spikes. The App and the web page have a long list of both individual foods and meals that are scored from A+ to C- in their ability to normalize your unique glucose spikes. Since Brett took the test, they have modified it to be based on a 1–10 score. It certainly makes for interesting reading, but there don't seem to be a lot of common or logical themes among food groups. A major concept to remember is that the individual food scores are just that—indicators of what happens when consuming that food in isolation, which we rarely do. So naturally, the scores significantly change when foods are combined—hence their meal-planning design function.

The goal of the suggestions is to have people stick to their suggested A+ meals or design meals with various foods that are A+ (the program also gives you a nutritional breakdown of the meal—carbs, protein, sugar, calories, etc.). It takes a bit of playing with the program to find certain meals and combinations that will appeal to your individual tastes, favourite foods, and cooking abilities. Once you have designed meals, you can save them for future reference.

Using the custom meal design, Brett found that a bagel with cream cheese and smoked salmon scored an A. Somewhat surprisingly, adding bacon to the meal kept it at an A, but adding a slice of whole wheat toast dropped the meal down to a B–! It also tabulates drinks' grades, although unfortunately doesn't include aspartame-sweetened drinks, which likely impact the microbiome big time, on the list. For example, Brett scored A+ with 86-proof whisky but B- for unsweetened grape juice (guess which one he will choose?).

The other useful aspect of DayTwo is that you get a thirty-minute phone consultation with a knowledgeable dietician who works with you to explain the results and helps you plan meals. This was really helpful in getting the full value of the analysis, including a very interesting discussion for Brett about what constitutes a "carbohydrate."

While the technology behind these plans is fascinating and fun, the big question is whether they work to improve health. Initial studies suggest that they may be more effective at weight loss and stabilizing glucose levels than traditional non-microbe-focused diet programs. DayTwo had many testimonials from people stating that it decreased weight and food cravings, increased energy, and improved metabolic measurements. Omri Casspi of the Memphis Grizzlies, the first NBA player from Israel (DayTwo's founders are Israeli, so much of the food list reflects that country's cuisine), used it and liked it so much that now the entire Israeli national basketball team is using it. Comparison studies with traditional diets are underway, which will truly determine the merit of this personalized diet concept and whether it will meaningfully help prevent obesity, diabetes, and metabolic diseases—and maintain healthy weights through all stages of life.

In Pursuit of Longevity, Trust Your Gut

At the start of this chapter, we saw that the microbial composition of extremely healthy older people resembled the gut microbiomes of healthy individuals many decades younger. Whether this is cause or effect remains unknown: Did the "ridiculously healthy" ninety-year-olds have diverse gut microbiomes because they were active and ate well? Or was their healthy aging predicated by the bacteria in their guts? Either way, there is a robust correlation between a healthy gut and healthy aging. Your microbiome is a biomarker of healthy aging.

We co-evolved with our microbes to mutual benefit. Minimizing disturbances to your gut's microbial ecosystem is thus essential to health and longevity. Disruptions can have devastating consequences, including increased risk for chronic GI diseases (e.g., Crohn's disease, ulcerative colitis, IBS) and metabolic disorders (e.g., type 2 diabetes, obesity). There are significant efforts underway to identify specific "core" gut microbiota that mediate specific disease mechanisms. Diet is a major and controllable environmental factor that influences microbial composition. We're looking at new ways to influence that with personalized diets, advanced probiotics and prebiotics that may help us re-establish "keystone" species. We are even examining methods of isolating probiotic strains from healthy older people or centenarians to develop therapies intended for healthier aging. We can envision a standardized cocktail of beneficial microbes obtained from healthy centenarians' feces.

Further establishing the safety, efficacy, and validity of FMTs could positively translate into broader usage with gut-related ailments including IBS, IBD, obesity, type 2 diabetes, and colorectal cancer. We might even reset the gut microbiota of an older person with microbes from a thirty-year-old to combat a specific disease or boost overall healthy aging. But given that gut microbiota and intestinal environments vary significantly between individuals, a lot more research is needed to precisely define and validate which microbes are considered "healthy" and "ideal" for everyone, and how to fill in the gaps where individual needs are concerned. In either case, research progress will hopefully allow for earlier detection and prevention of these diseases, so that our guts—and the rest of us—are happy for longer.

QUICK TIPS

- **It takes a village:** Maintain your gut's rich microbial diversity as much as possible throughout life. This includes taking antibiotics only when medically necessary, eating probiotic-rich foods (e.g., kimchi, sauerkraut, kefir, tempeh, yogurt, kombucha), and consuming prebiotic foods (e.g., asparagus, Jerusalem artichokes, bananas, oatmeal, red wine, honey, maple syrup, legumes; see more on these in Chapter 14). For a more targeted approach, consider one of the commercial products available to get a snapshot of your gut microbes and adjust diet accordingly.

- **Put away the yo-yo:** Cycling between diets is bad for your waistline and for your microbes, since your microbes likely remember their unhealthier composition when at a heavier weight and make sustained weight loss more difficult.
- **Poop-tacular:** If you or a loved one suffer from recurrent *C. difficile* infections (and potentially IBD), talk to your doctor about a fecal transfer instead of another round of antibiotics. Exciting clinical studies suggest that simply transferring feces from one person to another can potentially cure this deadly disease. But remember, *do not* try this at home!

8

Love Bugs: The Heart and the Microbiome

Your heart is a strong muscle roughly the size of the palm of your hand. Like the engine of a car, it's what keeps your body running. It has two main pumps: The first uses arteries to send oxygenated blood away from the heart to the rest of the body. The second uses veins to bring blood back to the heart and send it to the lungs to get more oxygen. The cardiovascular system spans your entire body: your heart, veins, arteries, and blood. Cardiovascular disease (CVD) happens when the engine and its pipes get blocked up. Plaque builds up in the arteries, which becomes a condition called atherosclerosis. If the plaque dislodges or too much of it accumulates, it can clog an artery and cut off critical blood supply to the body. If the blockage occurs in an artery that feeds the heart itself, it is a heart attack (a myocardial infarction, in medical-speak); if it occurs in an artery to the brain, it is a stroke. Blood clots and clogged arteries can also occur in legs (deep vein thrombosis) and lungs (pulmonary embolism).

One person dies from CVD every thirty-nine seconds in the United States. This disease is the leading cause of death worldwide. Every year, roughly one in three deaths globally (over seventeen million) is caused by a heart-engine that sputters out and stops pumping as it should. Although CVD is traditionally attributed to unhealthy diets and lifestyle, we now know there is a strong microbial component to the disease.

Age is by far the largest risk factor for CVD. The risk triples for every ten years lived, and there are indications that heart disease can be detected as early as in adolescence. Changes in the heart and blood vessels as you age increase the threat. The arteries become harder and less flexible, like a bike chain rusting over time, which is why blood pressure goes up with age. Blood vessels can become stiffer and the heart wall can thicken to assist with blood flow, i.e., more muscle to pump. The valves, which are like one-way doors that open and close to control blood flow, can also become thicker and stiffer. This leads to leaks or more resistance to pumping blood. Eighty-two percent of people who die of CVD are over the age of sixty-five, and the average age of death from CVD is eighty years old.

There's also a gender element to this rise, as older women are three to five times more likely to develop CVD than men. Cholesterol levels keep increasing in women until around sixty-five years of age (cholesterol buildup in the arteries is one of the things that inhibit the "engine" from pumping), whereas in men the levels plateau between age forty-five and fifty-five. Some believe this is related to hormones: Estrogen is protective of the heart, but estrogen levels decline following menopause (see Chapter 9 for more about menopause and microbes).

MYTH: Heart disease runs in my family, so there is nothing that I can do to prevent it.

FACT: Nearly all cardiovascular disease is preventable. Though people with a family history are at a higher risk, there are steps that you can take to immensely decrease your risk.

Microbes and CVD Prevention

As with many things we've already examined, when it comes to microbes and disease, how you live *now* affects your degree of risk later in life—it's never too late to have your eye on a heart-healthy lifestyle. Everyday decisions—like choosing a burger versus choosing a salad for lunch, or sticking to your new exercise program versus letting your gym membership slide—lead to effects on blood vessels that accumulate over time (think of all the burgers you've enjoyed over the years!). You're lucky if your parents gave you good genes, but when it comes to CVD most people are in control of their own luck.

It is estimated that nearly 90 percent of CVD is preventable, with only the remaining 10 percent due to family history and genetics. These preventable aspects include high blood pressure, which accounts for 13 percent of cardiovascular deaths; smoking, accounting for 9 percent (although if you quit smoking by age thirty this risk disappears later in life); obesity, accounting for 5 percent; poor diet; and excessive alcohol consumption. As challenging as it may seem, there is plenty you can do to prevent or delay CVD caused by these factors. Physical inactivity, which is defined as *fewer* than five moderate thirty-minute workouts per week, is thought to be the fourth leading factor for mortality worldwide, yet approximately 70 percent of the American population fails to meet this standard. Diets that are high in saturated fats, sugars, salt, and processed meats increase risk of CVD considerably, while diets such as the DASH diet, incorporating plenty of fruits, vegetables, fibre, and nuts, significantly lower the risk.

Given the major impact diet and exercise play in CVD, it's no surprise that the body's microbiota are involved in the disease's development and prevention. In patients with CVD, we see changes in the main types of intestinal microbiota: Increases in the *Firmicutes* and decreases in the *Bacteroides*. These are similar to changes seen in the microbiota associated with obesity (as discussed in Chapter 7), suggesting a shift towards more problematic "inflammatory" microbiota composition. As we've seen, the body defends itself with inflammation if any bacteria or bacterial cell-wall pieces, especially lipopolysaccharide (LPS), seep into the body. This defence—our body's response to infections—causes destruction of tissue, including damage to blood vessels and the heart. As we saw with the brain, low-grade inflammation is not healthy for the cardiovascular system.

CVD does not just affect the heart. Many people are unaware of how this condition extends throughout the body, including how the health of the mouth is related to CVD. We saw in Chapter 4 that the oral cavity is full of microbes, and that cavities and gum disease involve critical microbial pathogens. In gum disease, pockets of pus develop around the teeth, along with inflammation of blood vessels in the gum tissue. Scientists believe these gum changes enable periodontal pathogens to enter into the body's circulation through blood travelling via veins and arteries, where the pathogens encounter blood vessels and trigger more inflammation. This causes tissue damage, including the tissues of the cardiovascular system. Dental pathogens have been found in the heart valves of CVD patients. This is why if you have CVD, dentists will prescribe antibiotics prior to dental procedures to try to prevent microbes from directly entering the bloodstream during the oral surgery—although antibiotics will have additional, often unintended effects on the microbiota. Given that older adults are more at risk for periodontal disease and tooth loss, they are also more at risk for heart disease. Use this as additional motivation to keep brushing and flossing, and schedule regular cleanings with your dentist.

THE BIG FAT SURPRISE

There is a re-think underway regarding the role of saturated fat in CVD. Dr. Ancel Keys, a powerful scientific figure in the 1950s to 1970s, initially championed the concept that high saturated fat leads to high cholesterol and heart disease. He persuasively advocated for the American Heart Association and the broader scientific community to stress the need to decrease fat intake and make it public enemy number one. We all very much "took it to heart": Dietary fat intake in the US dropped by 11 percent since the 1970s, and grocery stores still stock many low-fat and non-fat foods.

But we had to eat something, preferably something that tastes good! So instead we increased our carbohydrate consumption—this includes adding sugar—by at least 25 percent. We now know that added sugar and refined carbohydrates also increase risk for CVD. Saturated fat is just one piece of the puzzle, and butter, meat, and cheese in moderation are part of a healthy diet. If you want the sordid details, check out *The Big Fat Surprise* by Nina Teicholz.

Here are the American Heart Association's current facts on fat:

- **LOVE IT:** Unsaturated fats lower rates of CVD. Healthy fats containing good HDL cholesterol, such as olive oil, beans and legumes, whole grains, fatty fish, and nuts act like a vacuum for the arteries. They remove bad LDL cholesterol and plaque buildup, and reduce risk of heart disease.
- **LIMIT IT:** Saturated fats are no longer forbidden but should be consumed in moderation. They occur naturally in many foods, including meat (e.g., fatty beef, lamb, pork, poultry with skin) and dairy products (e.g., cream, butter, cheese). It is recommended to aim for no more than 5 to 6 percent of daily calories from saturated fats.
- **LOSE IT:** Artificial trans fat, hydrogenated oils, and tropical oils, such as palm kernel oil, increase risk of CVD and raise bad cholesterol levels. Try to save commercially fried foods and baked goods for rare occasions.

Heart Chemistry, or the Science Behind CVD

Now let's get to the "meat of the matter" regarding microbes and CVD. We know that meat eaters, especially red meat consumers, have a higher CVD risk than vegetarians. In studies, germ-free—i.e., microbe-free—animals have a very low CVD risk, even if these animals are fed diets that increase the risk of CVD in animals that aren't germ-free; i.e., have a normal microbiota. Why is this? Meat and egg yolks are rich in two structurally related compounds, L-carnitine and choline, and these compounds make up about 2 percent of the Western diet. Through a series of reactions, the gut microbes convert these compounds into a molecule waste by-product called trimethylamine (TMA). TMA is then oxidized by liver enzymes to trimethylamine oxide (TMAO), which facilitates CVD.

This suggests that if we took the microbes out of the equation, TMAO wouldn't be formed (nothing would make the microbe by-product of digestion, TMA). By extension, the occurrence of CVD would decrease. Remarkably, this process has evidence in real life: Germ-free animals don't make TMAO, and in model systems they develop less CVD since they can't make TMA in the first place. But once they are colonized with microbes, they start producing TMA, then TMAO, and then get heart disease. In a study of four thousand people, high levels of TMAO equated directly to increased risk of high thrombotic events including stroke and heart attack, and TMAO levels could be used to predict the degree of risk. So the higher the TMAO levels, the higher the mortality from heart attacks.

MYTH: Red meat alone causes heart disease.

FACT: Red meat does not cause CVD by itself: It's how our microbes break down the red meat and create harmful by-products that leads to heart disease.

We interviewed Dr. Stanley Hazen, whose groundbreaking research has shown that TMA is a microbial product and TMAO is a major contributor to heart disease and stroke. A clinician and researcher in cardiovascular medicine at the Cleveland Clinic, Dr. Hazen explained that he "stumbled upon a connection to the gut microbiome" in the context of his work on heart disease and stroke, which ultimately led to these incredible revelations about TMA and TMAO. Working from a chemical background, Dr. Hazen's lab made their discovery around 2007 while searching large populations of patients for chemical signatures in plasma. They found a higher concentration of an unknown compound in patients with heart disease. After seeking out the rare and expensive tools needed to identify this compound, they discovered it to be TMAO, which was a known product of TMA, and yet Dr. Hazen could only locate descriptions of TMA found in putrefaction—i.e., bacteria growing on rotting things—when he began researching it separately. He figured that if there were enough bacteria to make it possible to see TMAO in plasma, then it must be coming from the body's microbial stronghold: the gut.

Dr. Hazen's team was shocked when he told them that they were switching gears and turning their focus to the gut. But their surprise soon turned to enthusiasm as the researchers realized that everything we eat is filtered through the intestinal tract, where its microbes digest, and help us absorb, our food. Dr. Hazen told us: "We like to think of ourselves as clean and sterile beings, but we aren't. We evolved with microbes through all of our eternity: from *Homo sapiens* onwards. It started to make total sense." Dr. Hazen explained that in his clinical work, he knew that genetic risk factors for heart disease only capture 10 to 15 percent of patient risk. The vast majority of risk is environmental. "Our gut microbes are the filter of our biggest environmental exposure: What we eat," Dr. Hazen clarified. "When two people eat the same diet, we can see one develop heart disease and the other not. A big chunk of this reason is the microbes." This was the enormous missing piece to understanding human physiology and disease risk.

Dr. Hazen's group took initial animal studies further by demonstrating similar effects in humans. He fed people eight-ounce steaks and then measured their TMAO levels, which increased soon after digesting this meat. He then put these people on antibiotics, which suppressed their TMAO production because the key microbes that make TMA were killed off. After he stopped the course of antibiotics, the microbes and TMAO production increased to previous levels. But his coolest experiment involved convincing a vegan of over five years to eat a steak! Though he promised there was no coercion involved for the sake of science:

> In our early studies, it was important to provide some food when giving study participants a capsule. For those who agreed, we served them a steak. In one test, someone ate a filet mignon prepared on a George Foreman grill in my office—not too untypical during this time. But this

> person's blood and urine sample stood out like a sore thumb. When we did a follow-up to check for any genetic diseases, this person told us he/she had only eaten steak three times, ever—I had no idea this particular person was a vegan! This was a game-changing moment that shifted our gears towards considering differences between vegans, vegetarians, and omnivores.

As one would predict, the post-steak TMAO levels of the vegan were much lower than those of regular meat eaters—because the vegan had fewer microbes that could make precursor TMA to begin with. Outside of Dr. Hazen's discovery, further studies have also shown that there are higher levels of TMAO in omnivores than in vegans and vegetarians—which confirms the connection between certain animal products such as red meat and this microbial menace of the heart.

Red meat is not the only source of these compounds: Choline, lecithin, and carnitine supplements can also give rise to the body's production of TMAO. Drinking multiple cans of some energy drinks, which can include supplemental carnitine, may be equivalent to eating multiple steaks' worth of carnitine a day. Research has shown that feeding mice diets rich in carnitine or choline can cause a tenfold increase in TMAO, as well as an increase in the microbes that form TMA, which then leads to TMAO.

With growing interest in microbes and CVD, researchers discovered that TMAO is what leads to the "clogged pipe" of atherosclerosis by suppressing reverse cholesterol transport from peripheral tissues back to the liver. It also affects platelet activity, which can lead to more clotting of blood vessels (thrombosis). Both of these conditions obstruct blood flow and cause heart disease.

Compelling evidence from fecal-transfer experiments further demonstrates that gut microbes are involved in CVD. When fecal matter from a mouse line that makes high levels of TMAO was transferred to a breed of mice with low TMAO levels, the recipients' TMAO levels went up, as did the microbes with the enzyme that makes TMA. By simply swapping feces (and microbes), it would be possible to alter the likelihood of CVD! This would have been an unthinkable concept a decade ago, but now scientists are hard at work to see exactly how microbes are involved in heart attacks and strokes. In the future, perhaps such measures could be taken in humans, to similar degrees of success.

SWEATING THE MICROBES

Exercise plays a key role in protection against CVD, and new data suggests this may occur in part through changes to the microbiota. When mice were fed normal or high-fat diets, both lean and fat mice that then "exercised" (unlimited access to a running wheel) did not have the arterial damage seen in those that didn't exercise (no wheel). There were also more beneficial shifts in the mice's microbiome following exercise, including increases in *Faecalibacterium prausnitzii*, which protects against obesity and IBD, since it decreases inflammation. Researchers found that exercise also improved the integrity of the gut (meaning less leakage of LPS and resulting inflammation) and, regardless of diet, it was uniquely beneficial to the microbiome. Of particular relevance here is another key factor: Even exercise in older animals reduced inflammatory markers and increased intestinal health.

Further interesting studies like this suggest how microbes affect endurance as well (we may be on the cusp of a bug-doping trend; see Chapter 12). Earlier studies (with different animal welfare ethics standards) found that when mice were made to swim to exhaustion, the mice colonized with microbes fared much better than germ-free mice. Animals colonized with microbes also had increased muscle mass and increased antioxidant activity, which helps protect against intense exercise-induced oxidative damage (read: sore muscles).

What does this mean for us? Given that 70 percent of the population is considered inactive, these studies suggest boosting our activity may affect CVD incidence. A study out of Ireland sheds some light on similar effects in humans. The research group studied a professional Irish rugby team during their pre-season camp and compared them to normal people. They found that the athletes had a much higher diversity in their microbes—which is generally good for health. Controls with low BMI and the athletes also had higher levels of *Akkermansia*, a beneficial bacterium that increases short-chain fatty acids; this in contrast to low levels of this microbe found in those with obesity and metabolic diseases. What these studies seem to point to is that exercise, like diet, does directly, and beneficially, influence the microbiota.

Drugging the Bugs

How can we use these discoveries to decrease our risk of CVD? One might think that targeting the human liver enzymes that convert TMA into TMAO with inhibitors might reduce CVD. But when we try this, we raise the level of the precursor TMA, which smells like rotting fish. Not a good side effect for a drug! Several studies that genetically "knocked down" the liver enzyme resulted in mice that stink like rotten fish—but they were protected from CVD. Targeting liver enzymes as such is difficult pharmacologically in humans—all the compounds were toxic, in addition to the unattractive smell. For these reasons, scientists have been unable to target the key liver enzyme in TMAO formation.

Perhaps we could go after the microbes directly? Hazen's group has done this in animals. They identified a naturally occurring compound (3,3-dimethyl-1-butanol, or DMB) found in grape seeds, red wine, and cold-pressed extra virgin olive oils (think of the Mediterranean diet); this compound inhibits the bacterial enzyme (called a lyase) that initially forms TMA from choline and L-carnitine. Amazingly, when scientists fed the inhibitor DMB to mice, they observed decreased TMAO production and atherosclerosis. DMB didn't kill the microbes—which is good because that would put pressure on them to evolve and increase their resistance (much like how resistance to antibiotics develops). Although this inhibitor has only been tested in mice and the team is looking for even better inhibitors, the results certainly suggest that drugging the bugs may soon become a viable strategy to decrease risk of CVD.

Cardiologists and physicians are beginning to incorporate some of these findings into patient care. TMAO testing is now available, and some are using TMAO as a clinical diagnostic tool to recognize previously unidentified individuals who are at risk for CVD, and those who may be overlooked by traditional tests such as markers for cholesterol. Dr. Hazen advised that someone with a high TMAO level should take a progressive approach to CVD prevention through weight reduction, improvements to diet (such as cutting down on animal products), and watching traditional markers of blood pressure and cholesterol. As always, there is a need to tread carefully regarding modifying TMA or TMAO levels, Dr. Hazen cautioned, given the lack of reliable proof of efficacy and long-term safety. "TMAO is just the tip of the iceberg. Dozens and dozens of other connections between gut microbes and cardio-metabolic processes exist. It's going to get a lot more complicated before it gets simpler." Dr. Hazen expressed his belief in rigorous scientific trials to ensure that we do no harm.

NO WHINING ABOUT RED WINE

Alcohol over-usage is perhaps the second-best known risk factor in CVD after meat, but in reality its consumption is a double-edged sword when it comes to the heart. There is no doubt that high levels of alcohol consumption increase the risk of heart attack. People who had heart attacks were asked if they had a drink within the hour before the attack; those who had consumed hard liquor drinks (e.g., gin, vodka, whiskey) had increased their risk more than if they had consumed beer or wine, but the risk with beer or wine was still greater than for those who did not have an alcoholic beverage. Binge drinking is defined by the National Institute on Alcohol Abuse and Alcoholism as a pattern of drinking that brings blood alcohol concentration levels to 0.08 g/dL—typically occurring after four drinks for women and five drinks for men in about two hours. Binge drinkers are 72 percent more likely to have a heart attack than those who consume alcohol moderately—defined as up to one drink per day for women, and up to two per day for men. Even infrequent binge drinkers can have increased gut permeability and higher levels of inflammatory LPS and bacterial DNA. This indicates microbes in their blood serum following the binge could contribute to general inflammation.

The good news is that a glass of red wine a day is protective for CVD, similar to its health-promoting work in the gut, as discussed in Chapter 7. (We think it's the best part of the MIND, Mediterranean, and DASH diets.) Although we previously thought it was the antioxidants in red wine that helped protect our arteries from inflammation damage and CVD, no such mechanism had been found. Instead, it's microbes that are causing us to rethink how red wine might be beneficial.

Red wine, grape seeds, and olive oil all contain a compound called DMB, which inhibits the microbial enzyme that converts meat products to TMA. This ultimately decreases TMAO levels and atherosclerosis in animals. The benefits of resveratrol, a polyphenolic compound found in the skin of grapes and other berries (and in red wine), may be similar. Researchers found that resveratrol causes a reduction in TMAO levels in mice—but it also causes a restructuring of all the intestinal microbiota, most importantly the decrease of the microbes that produce TMA. Although it is early days for trials on humans, it is suggested that red-wine polyphenols (such as resveratrol) favourably change the microbiota over a month.

All this definitely calls for a glass of red wine and a toast to our microbes!

The Future

Given the exciting developments described in this chapter, CVD prevention will undergo a major revolution in the next few years. We are now learning how microbes are beneficially affected by lifestyle (e.g., exercise) and diet (e.g., red wine). The most promising leads involve targeting microbial enzymes to prevent disease with new drugs that will stand a much better chance of minimal side effects since the pathway they are targeting is only found in microbes, rather than in humans. One possibility is designing probiotics or prebiotics to outcompete these unwanted organisms and their genes. Moreover, preventive analysis of an individual's microbiome can deliver an indication of the level of microbes present that contribute to CVD. This could provide us with specific diet modifications to target and decrease these microbes before a heart attack or stroke occurs—we can already analyze the microbes' presence, but the exact dietary changes needed to correct them are less clear. Specific microbial interventions promise new ways to target the world's number one killer in the future.

QUICK TIPS

- **Eat and drink to your heart's content:** Incorporate plenty of fruits, vegetables, whole grains, fibres, and nuts into your diet. Keep your heart and microbes happy by limiting intake of saturated fats, added sugars, salt, and red meat. Red wine can inhibit TMAO production and help protect the arteries. Cold-pressed extra virgin olive oils and foods rich in antioxidants (e.g., dark chocolate, green tea) and good HDL cholesterol (e.g., nuts, avocados) in moderation can also protect the heart.

- **Smile with your whole heart:** Regular brushing, flossing, and dental check-ups can help keep away heart disease. Don't rush when you brush to get all those deep cavities where microbes can settle down!
- **Get your blood pumping:** Regular exercise is essential to the health of your heart and your microbes. Pursue whatever heart-pumping activities you enjoy—walking, hiking, swimming, cycling, gardening, even dancing. It is recommended to get at least thirty minutes of moderate activity most or all days of the week, and this doesn't have to all be done at once. Ten-minute blocks of exercise are just as beneficial (if not more so, since you are sitting less continuously the rest of the day).

9

Females Are Not Small Males: Menopause and the Vaginal Microbiome

As an avid triathlete-turned-runner, Jessica is constantly seeking scientific information on training, nutrition, and recovery. Her constant frustration, however, is that the science is most often based on college-aged males. In fact, until the 1980s it was generally assumed that physiological responses to exercise did not differ between the sexes. Yet we cannot simply "shrink it and pink it," given the very fundamental physical and hormonal differences between males and females—when it comes to exercise *and* the microbiome.

So far, this book has been fairly gender-neutral, and for the most part the microbiome can be discussed as a constant across men and women. This chapter focuses on hormonal and microbial processes that most women experience, and so we have used the term "women" throughout for simplicity, although we recognize that these processes are not experienced by all women, and are not exclusive to people who identify as women. In general, there are fundamental physiological and hormonal differences between males and females that are reflected in their unique microbiomes. Women have periods, for example, and it's okay to talk about it. The vaginal microbiome changes throughout a woman's lifetime, especially as she ages. This chapter discusses specific microbial signatures of the vagina, changes that can occur during menopause, and the role of the gut microbiome in moderating estrogen levels. It also dives into emerging understanding of the urinary tract microbiome.

Habits of a Healthy Vagina

Females are not the same as males in a strictly biological sense. The vagina—the elastic, muscular part of the female genital tract that extends from the vulva to the cervix—is unique to women and critical to sexual intercourse, childbirth, and the menstrual flows it channels. It is interrelated to other organs in the female pelvis, including the urinary bladder and the bowel, which lie in close proximity. The vagina harbours a remarkably characteristic microbiome that forms a mutually beneficial relationship with the rest of the body and has a major impact on both health and disease. Like the gut, the normal vaginal microbiome forms a complex ecosystem of more than two hundred bacterial species. Its composition is affected by numerous factors, including ethnic background, environmental and behavioural factors, and stage of life—prepuberty, adolescence, reproductive years, and menopause.

Because it is relatively easy to access, the vaginal microbiome has been cultured for many years. Traditionally, it was thought to be dominated by *Lactobacillus* species, which produce lactic acid and give the vagina an acidic pH (3.5 to 4.5)—providing strong protection against pathogenic microbes. Lactobacilli also produce other antimicrobial products (bacteriocins) that specifically lyse (disintegrate) and kill other bacteria, again protecting the vagina from pathogens. Interestingly, humans are one of the few mammals that have *Lactobacillus*-rich vaginal microbiomes with a low pH.

We significantly advanced our understanding of the vaginal microbiome when culture-independent sequencing methods came of age, allowing us to pick up microbes that weren't as easily cultured as *Lactobacillus*. Scientists identified five general types of vaginal microbiomes in normal healthy women, each with a unique microbial signature. The majority of women (one study cites 73 percent) have one dominant *Lactobacillus*: either *L. crispatus* (found in 26 percent of women in that study), *L. iners* (predominant, in 34 percent of women), *L. gasseri*, and *L. jensenii*. Even in the microbiomes in which one particular *Lactobacillus* is dominant, it is common to find the others present, just in lower numbers. The fifth type of vaginal microbiome identified lacks a dominant *Lactobacillus* and is populated by a diverse collection of strict and facultative anaerobes (meaning that they are either killed by oxygen or can tolerate oxygen but prefer growing without it). Some studies indicate that different *Lactobacillus* species can lower the vaginal pH to different levels. White and Asian women are more likely to have a lactobacilli-dominant microbiome compared to Hispanic and Black/African American women. Interestingly, Black/African American and Hispanic women have higher average vaginal pHs, at 4.7 and 5.0 respectively, compared to Asian (4.4) and White (4.0) women. Up to 40 percent of Black/African American and Hispanic women have a non-lactobacilli dominant microbiome. To give you an idea of how complex these communities are, non-lactobacilli microbiomes include members of the genera *Atopobium, Corynebacterium, Anaerococcus, Peptoniphilus, Prevotella, Gardnerella, Sneathia, Eggerthella, Mobiluncus,* and *Finegoldia*, among many others. The finding that healthy women can

MYTH: The vagina naturally smells bad.

FACT: It's normal for the vagina to have a slight odour. This can vary depending upon menstrual cycle, sexual activities, and sweat. Distinct causes of abnormal odours include bacterial vaginosis, which occurs when the delicate balance between "good" and "bad" vaginal bacteria is upset and causes an overgrowth of anaerobic bacteria. It is different from yeast infections, which are caused by a fungus (*Candida*). A yeast infection generally itches; has a thick, white discharge; and doesn't smell. Bacterial vaginosis, in comparison, has a thin white, grey, or green discharge; a burning sensation when you pee; and a fishy smell. A doctor may prescribe a course of antibiotics for bacterial vaginosis, while antifungal medications are generally used to treat yeast infections.

have a non-lactobacilli microbiome and a vaginal pH greater than 4.5 has really challenged the conventional wisdom about what was previously thought to be needed for vaginal health.

Vaginas vary significantly from woman to woman. Periods also range widely: While the average menstrual cycle is twenty-eight days long, it can last anywhere from twenty-one to thirty-five days in adults. Menses cause significant changes in the vagina, which are reflected by transient changes in the vaginal microbiome. In general, the lactobacilli decrease and there is an increase in other microbes that can be associated with vaginal infections. Sexual intercourse also disrupts the vaginal microbiome, decreasing the lactobacilli and their protective effects. This can lead to a more dysbiotic composition, which is one of many factors that can explain why some women can get urinary tract infectious (UTIs) following intercourse—with less of the protective, acidic environment, more harmful microbes can enter the body. In some women, bacteria can also enter the bladder during intercourse.

Like the rest of the body, the vagina and vulva change significantly with time, as of course does the vaginal microbiome. Prior to puberty, females have a polymicrobial community rich in Gram-negative anaerobic bacteria and low in lactobacilli and *Gardnerella vaginalis*. Following puberty and incumbent hormonal changes, women develop one of the five vaginal microbiomes described on pages 159–60. Postmenopausal women have lower levels of estrogen, which reverts the microbiome to one that resembles a pre-puberty composition.

Maintenance of the normal vaginal microbiome is quite important in preventing disease. Bacterial vaginosis (BV) is the most common vaginal infection in reproductive-age women. It affects between 10 and 15 percent of women and results in millions of healthcare visits. The risk of BV is increased by changes to the vaginal environment including menstrual blood, a new sexual partner, and vaginal douching. In the United States, an estimated 20 to 40 percent of women aged fifteen to forty-four years old douche. Most of the time, BV does not cause serious health problems, but it can lead to genital tract and pregnancy complications, pelvic inflammatory disease, and increased susceptibility to sexually transmitted diseases. BV is often accompanied by vaginal discharge and odour.

Not surprisingly, BV is associated with dysbiosis of the vaginal microbiome. It features a decrease in the lactobacilli and an overgrowth of a polymicrobial community consisting of strict or facultative anaerobes. Although we saw on page 161 that about one-quarter of women have microbes similar to this kind of microbiome and are normally healthy, we just don't understand why at this time. Women with BV tend to have thick biofilms (large clusters of adherent microbes) growing on the surface of vaginal cells, especially *Gardnerella vaginalis* and *Atopobium vaginae*. However, there does not appear to be a single microbial causative agent. This leads to the theory that BV is linked to changes in the overall vaginal community—which, as we have seen, seems to be a recurring theme when discussing dysbiotic microbiota. Biofilms are notoriously resistant to antibiotics as they have trouble penetrating the microbial layers. Treatment of BV with antibiotics such as metronidazole is often unsuccessful, resulting in recurrence rates of about 30 percent after three months, and 60 percent after six months. In most women, BV symptoms usually resolve without interventions.

MYTH: Douching—spraying the inside of your vagina with a mixture of water and vinegar or a multitude of other "cleansers"—is an effective way to clean your vagina.

FACT: You do not need to douche to clean your vagina. The body naturally flushes out and cleans the vagina, whereas douching disrupts the normal flora and pH of the vagina. Most doctors recommend that women do not douche, as studies have not found any health benefits. Douching is linked to health problems including bacterial vaginosis, pelvic inflammatory disease, pregnancy complications, sexually transmitted infections, and vaginal irritation—all likely affected by the microbial disruptions it causes.

Full Stop: Microbes and Periods

There are two major instances in a woman's life that involve profound bodily changes linked to reproduction: when menstruation begins and when it stops. These are the bookends of a woman's ability to bear children. Women are born with a finite number of eggs, which are stored in the ovaries. The ovaries also make the hormones estrogen and progesterone, which modulate menstruation and ovulation. At puberty, the ovaries begin releasing those eggs once a month, and if no embryo is conceived the lining of the uterus is shed during menstruation. (The word "menstruation" comes from the Latin *mensis*, meaning "month.") When a woman's full store of eggs has been released, menstruation stops. This point is referred to as menopause (think: menstruation + pause) and usually occurs around age forty-eight to fifty-five in the United States (fifty-one on average), after a woman has gone without a period for twelve consecutive months. The years following the point of menopause are often referred to as "post-menopause."

Menopause is a normal phase of life. It is a part of aging, not a disease. Experiences of menopause are more and more common given the increasing life expectancy of women worldwide. Menopause is associated with lower levels of reproductive hormones, given the reduced functioning of the ovaries, especially estrogen. Overall lower reproductive hormones are associated with increased risk for osteoporosis, bone fractures, and loss of muscle (sarcopenia; discussed in Chapter 12). Low estrogen can result in body temperature swings (hot flashes and sweating), psychological changes (mood shifts, irritability, and depression), insomnia, vaginal dryness, and bladder control problems. Lower androgen levels (male hormones) can contribute to reduced sex drive. There is great variation in how women experience menopause, and not all women undergo all of these symptoms. For those who encounter severe symptoms that affect quality of life, hormone replacement therapy (HRT) is designed to help. HRT is basically a way to supplement estrogen. Hormone replacement can bring risks, however, so talk to your doctor if you are considering treatment.

The human gut microbiome is increasingly thought of as an endocrine organ: one that produces (or affects) hormones, which affect body function. Although the ovaries are the primary producer of estrogen prior to menopause, later in life other cells of the body can still produce estrogen. This includes cells in the gut. Germ-free rats secrete very low levels of estrogen compared to microbially colonized rats—levels so low that they do not have menstruation cycles and have trouble reproducing. When the germ-free rats were recolonized with microbes, their estrogen levels increased, and reproduction and menstruation cycles were restored. The gut microbiome clearly affected estrogen levels.

Throughout life, female bodies strive for balance in estrogen levels. Too little estrogen can cause headaches, hot flashes, night sweats, vaginal dryness, and other menopausal symptoms described previously, which can affect quality of life. On the flip side, having too much estrogen can also wreak havoc on the body with irregular periods, bloating, weight gain, headaches, anxiety, decreased sex drive, and increased symptoms of premenstrual syndrome. We want to keep hormonal levels balanced, like an equally weighted seesaw, to avoid these health symptoms. Because gut microbes have specific enzymes (*beta-glucuronidases*) that can cleave precursors of estrogen, called *phytoestrogens*, estrogen can be produced in the gut. This presents a novel avenue to potentially help moderate estrogen levels in menopausal women, including through diet.

Phytoestrogens are found in some plants: Isoflavones in soy products (milk, bean curd, sprouts) and lignans (a compound abundant in seeds, particularly flax, berries, fruits, vegetables, and whole grains). The gut microbiome deconjugates (breaks down and metabolizes) these estrogen precursors to produce active forms of estrogen that can then be absorbed by the body. Once in the body, they can act like weak hormones and may reduce the risk of some diseases associated with low estrogen such as osteoporosis.

Studies show that strict vegetarians have increased conjugated estrogens in their feces. This indicates that the microbial enzymes in their guts that normally deconjugate these compounds to release active estrogen into the body are not present, or not active to the same extent, compared to omnivores. Studies of women on a Western diet (high fat, low fibre) compared to vegetarian women showed that vegetarians had 15 to 20 percent less estrogen in their circulation and triple the conjugated estrogen in their feces. In a study of sixty menopausal women, researchers found that women with diverse microbiomes had less estrogen and more breakdown products than women who had a less diverse microbiome. We don't yet know exactly why this is, but it is certainly a topic for further studies.

The concept that the gut microbiome might alter estrogen levels is an exciting one. Perhaps we can design HRT to work even better by incorporating a microbial approach to modulate estrogen levels during menopause. This could decrease risk for menopausal conditions including obesity, glucose intolerance, and metabolic diseases. Stay tuned for additional scientific research focused on the relationship between estrogen levels and microbial enzymes.

Menopausal Microbes

Given the significant changes in the vaginal environment during menopause, it is not surprising to see more direct changes in menopausal and post-menopausal vaginal microbiomes—though unlike premenopausal vaginal microbiomes where there are monthly shifts in the community, the post-menopausal vaginal microbiome seems remarkably stable. During menopause (the point when periods stop), the microbiome is still dominated by lactobacilli (although to a lesser degree, in total numbers), including *L. crispatus* and *L. iners*, as well as *Gardnerella vaginalis* and *Prevotella*; and a lower abundance of *Candida, Mobiluncus, Staphylococcus, Bifidobacterium*, and *Gemella*. In the years following menopause (referred to as "post-menopause"), there is an overall decrease in lactobacilli and replacement with other microbes. The lactobacilli can degrade glycogen into organic acids, especially lactic acid, which keeps the vagina acidic. After menopause there is less lactobacilli and less lactic acid production. This then increases the vaginal pH, which some believe may help to explain an increase in risk for UTIs and gynecological infections among older women.

Estrogen keeps the vaginal surface healthy by triggering the deposition of glycogen (a sugar) on the vaginal mucosal surface. Like plants in need of regular fertilizer, these vaginal microbes, including lactobacilli, need glycogens for sustenance. Lack of estrogen causes the genitourinary syndrome of menopause (GSM)—also termed vulvovaginal atrophy, vaginal atrophy, and atrophic vaginitis. This leads to a thinning of the vaginal wall, which means less glycogen deposition and fewer of the beneficial resident microbes. Symptoms of GSM include dryness, itching, burning, urgency to pass urine and/or urinary leakage, lack of lubrication and bleeding during intercourse, vaginal discharge, susceptibility to sexually transmitted infections, and inflammation of the vagina. Up to half of all post-menopausal Western women experience these symptoms, with vaginal dryness being the biggest complaint.

In a Swedish study that compared twenty fertile (pre-menopausal) women to twenty post-menopausal women, fertile women had far more diversity in the subspecies of lactobacilli. Another study confirmed the correlation between glycogen levels and *Lactobacillus* species: Post-menopausal women have decreased glycogen and lactobacilli. There are strong correlations between vaginal health and lactobacilli—and inverse correlations between *Gardnerella* and *Atopobium*, two biofilm-forming species also seen with BV. Generally, but not always, GSM is associated with lower levels of lactobacilli. One study of post-menopausal women divided them into groups based on vaginal dryness. In the group that had no to mild vaginal dryness, the women were rich in *Lactobacillus* species and had low bacterial diversity, both positive markers of vaginal health. In the group that had more severe vaginal dryness, the women were colonized with fewer lactobacilli, and had more bacterial diversity, including *Prevotella, Porphyromonas, Peptoniphilus*, and *Bacillus*. Individuals with more severe signs of GSM tend to have higher vaginal microbiota diversity that is not dominated by lactobacilli. In a sense, vaginal health is promoted through a dysbiotic population—contrary to nearly every other finding in this book.

Somehow, we need to restore high lactobacilli levels while decreasing the bacterial diversity for post-menopausal vaginal health.

If lack of estrogen causes these issues, wouldn't adding it back do the trick? Estrogen levels and healthy microbes do seem to go hand in hand. In one study, 44 percent of post-menopausal women *not* on HRT lacked lactobacilli compared to only 6.9 percent of post-menopausal women on HRT. All routes of estrogen administration (oral, injected, topical, or vaginal) can improve vaginal health and microbiota. Lactobacilli species normally seen before menopause seem to return to similar numbers, to the benefit of vaginal health. In a study of women treated with estrogen for three months, 20 percent of controls compared to 80 percent of women on estrogen reported improvements to vaginal dryness and irritation. The group also saw increases in lactobacilli and lowered vaginal pH. There were no changes in sexual complaints based upon placebo or estrogen. This suggests that low dose estrogen therapy may work well for those with GSM to bring the lactobacilli back and decrease diversity and vaginal pH. Yet there are also perfectly healthy women who have a high vaginal pH of 6.6, extremely little lactobacilli, and high microbial diversity. Like the gut, every vagina is unique: What a healthy microbiome looks like in one woman can differ markedly from other healthy women, which means that broad generalizations are not universally applicable. It's also worth keeping in mind that high doses of estrogen in post-menopausal women can increase risk for endometrial and breast cancer. The potential dangers of HRT are higher for certain groups of women, including those already at risk for breast cancer, coronary artery disease, stroke, or active liver disease. Estrogen supplementation can be given as low dose vaginal supplements in the form of cream, tablets, or estrogen eluding rings (vaginal inserts).

Given the importance of the vaginal microbiome, there is significant promise and early progress using microbes (especially lactobacilli) as probiotic treatments. These can be delivered orally or vaginally. Vaginal probiotics presumably work in several ways: Reconstituting the normal microbiome, decreasing acidity, out-competing microbial pathogens, and enhancing the epithelial barrier and mucosal immunity. There are several reports of both oral and vaginal delivery of *Lactobacillus* in post-menopausal women suggesting that probiotics, compared to controls, may improve vaginal symptoms and health. A trial that tested the effect of vaginal delivery of low-dose estriol (estrogen) and *Lactobacillus acidophilus* in post-menopausal women found that it improved vulvovaginal symptoms compared to controls. The concept of delivering both estrogen and a healthy normal flora microbe, *Lactobacillus*, is particularly appealing given that both of these tend to decrease post-menopause. However, as noted earlier, there are health risks to HRT, including heart disease, stroke, blood clots, and breast cancer.

Probiotics can have marked effects both before and after menopause. One study in premenopausal women showed that after using a vaginally delivered probiotic for six months, the vaginal microbiome became richer in *Lactobacillus* compared to the controls. It seems the generic oral lactobacilli probiotics, which are not isolated from the vagina, may not function as well. This is likely because they are not normal vaginal microbes and may actually disrupt the communities. In Chapter 14, we discuss in detail how to find out which probiotics have undergone rigorous clinical testing.

HOW SMART IS JANE?

Want to know your own vaginal microbes better? In late 2017, uBiome launched the first vaginal microbiome/pathogen detection kit: SmartJane. Like the fecal analysis used to figure out the gut microbiome, SmartJane involves a simple vaginal swab taken at home and mailed back to uBiome for processing. DNA analysis is done to detect a few specific microbes associated with health and disease—unlike gut microbiome tests, it does not sequence them all. These microbes include fourteen of the most common high-risk strains of human papillomavirus (HPV) associated with cervical cancer, including HPV-16 and HPV-18. HPV is a sexually transmitted virus associated with cervical cancer, and there is now an adolescent vaccine widely administered to protect against this virus and prevent cervical cancer. SmartJane detects five low-risk HPV strains associated with genital warts—including HPV-6 and HPV-11—that account for the vast majority (90 percent) of genital warts, although rarely do they turn into cancer. SmartJane also tests for four other common sexually transmitted infections caused by bacteria: chlamydia, gonorrhea, syphilis, and *Mycoplasma genitalium*. Finally, the test identifies twenty-three common vaginal microbes, including many lactobacilli. It compares these microbial levels to a healthy reference range. According to their website, the test is friendly to trans and non-binary gendered patients.

Unlike fecal tests that can be purchased directly by individuals, this test can only be ordered by a healthcare provider. It marks the first time that a microbiome test is being coupled with healthcare providers and covered by health insurance companies. SmartJane is billed as a way to measure vaginal health. It is *not* a test for cervical cancer (although HPV testing in the kit may give a good indication of cancer risk) and does not replace Pap smears (which do test directly for cervical cancer) or routine vaginal exams. The test is designed to inform patients and their doctors about vaginal flora balance—although they are testing for only some of the many vaginal flora microbes. Knowing the flora can provide information about bacterial vaginosis associated with microbial dysbiosis.

SmartJane is not meant to be a DIY kit: Because it requires a physician's request, it falls in the realm of medical tests, and therefore into the realm of reimbursement from healthcare insurers rather than out-of-pocket payment. It will be interesting to follow SmartJane's accessibility and utility as microbiome testing enters into the healthcare arena.

The Bladder Fallacy

Scientists traditionally thought that urine and the bladder that holds it prior to urination were sterile. This thinking goes back to the mid-1800s and is based on the fact that a vial of sealed urine does not turn cloudy. Dr. Lynn Stothers, a professor of urology at the University of British Columbia, a research scientist, and a practicing urologist, explained this history of urology: "The bladder was long thought to be sterile. Somehow the urethra was like the gatekeeper between the vagina and the bladder. But later we found that some people have 'asymptomatic' bacteriuria [i.e., bacteria can be in the urine but there may not be any symptoms]." Much of the research on the vagina was conducted under the impression that there were no organisms in the bladder. If any microbes were detected from the bladder, they were explained away as likely contaminants from the vagina or the result of a UTI. The Human Microbiome Project (data collected from 2008 to 2013) did not sample the bladder for sequencing, as it was considered sterile at that time. This dogma persisted until very recently. Sequence-based analysis has now revealed that there are indeed microbial signatures in the bladder. Dr. Stothers admitted: "This discovery generated a lot of surprise. Clinically, we know that approximately 20 percent of women have bacteria growing in their urine and are asymptomatic. How can that be? It has created clinical confusion that persists to date." Many clinicians, still under the assumption that the bladder and urine should be sterile to be healthy, unnecessarily and inappropriately prescribe antibiotics to rid the body of bacteria that are not causing any symptoms.

MYTH: Urine is sterile.

FACT: There are numerous bacteria present in urine—and not just in individuals with a UTI. Bacteria are present in the pee of healthy individuals too.

Extended efforts are underway to culture microbes from urine. "We are trying to move into the idea that the bladder is not always a sterile environment," Dr. Stothers explained. "The trouble is, however, that it is difficult to obtain suitable specimens from the bladder. How can we collect urine? It is easiest and most practical for people to pee into a cup and then study the voided urine. To get urine directly from the bladder is much more difficult. We can pass a catheter (a hollow tube) up the urethra or put a needle through the abdomen to aspirate urine from a full bladder, but either way urine is acquired in an invasive manner. This is a barrier to research." A lot of work has been done to compare voided samples to urine obtained directly from the bladder. Under the right conditions (enhanced quantitative urine culture), scientists have found that there are indeed microbes in the bladder. There are greater than 100,000 viable microbes per millilitre of urine, which consists of a variety of microbes, many—including *Lactobacillus* and *Gardnerella*—similar to those in the vagina. This number of microbes has traditionally been used as a threshold for determining bladder infections, but we now realize that even numbers much larger than this do not necessarily represent infections.

Do microbes in the bladder and urine play a role in health and disease? Like so many examples in this book, there are hints that they do, but we don't know much about the mechanisms yet. The urinary microbiome, for example, is different in female overactive bladder patients compared to controls. Those with an overactive bladder have more diverse microbes and fewer lactobacilli—so it is thought, but again not proven, that microbes like the lactobacilli are protective against pathogens that can infect the urinary tract. The microbiome of urgency urinary incontinence patients is different still, with more *Gardnerella* and diverse species than continent controls. Dr. Stothers clarified that until we can routinely collect clinical samples that provide a specific profile of organisms in the microbiome, it will be difficult to translate scientific discoveries into medical practices: "We are now just barely dipping our toes into the world of the urinary microbiome. The tipping point into clinical medicine is when we can get urinary microbiomes on a day-to-day basis. At the moment, clinicians can only get a standard urine culture. We need to be able to see reports with the urinary microbiome." To do this, we need to convince clinicians who are still skeptical that a urinary microbiome even exists.

MYTH: Cranberry juice can cure UTIs.

FACT: *Prevention* of a UTI is different from treating an active symptomatic infection. The body of evidence supports cranberries in the prevention—but not "cure"—of UTIs. Scientists originally thought that the acidity in cranberry juice decreased vaginal pH, which could protect against pathogens. More recent data indicates that components in the juice bind to—and plug up—surface structures (called pili) on the microbial pathogen's surface, therefore blocking its ability to adhere to host cells. Numerous clinical studies have tested the ability of cranberry juice to block (prevent) UTIs. Because some studies suggest that cranberry products may reduce repeated infections in women, some medical professionals may suggest the use of cranberry supplementation in the prevention or treatment of a UTI. Dr. Stothers recommended it as "a 'low risk' prevention strategy in patients with recurrent UTIs with potential upside (as opposed to chronic antibiotic use)."

Dr. Stothers was hopeful that advancing knowledge of the urinary microbiome could be harnessed to better identify and treat urinary conditions. "First of all, we can potentially influence our urinary microbiomes through diet. We know that compounds from foods we eat, such as cranberries, end up in our urine. Hopefully we can use things like diet to more precisely alter the urinary microbiome to prevent UTIs." UTIs also involve a variety of microbial pathogens, especially *E. coli*, that infect the kidneys, urethra, or bladder. A healthy vaginal microbiome may serve a protective role by outcompeting pathogens.

Could we use probiotics to increase this protection? There is not strong enough data to make this claim. In a review of nine published studies (735 participants total) that tested the effectiveness of probiotics to prevent UTIs, reviewers concluded that the currently available evidence indicates no significant reduction in risk of recurrent bacterial UTIs.

Dr. Stothers also mentioned important intersections between the vaginal and urinary microbiome, and how these two regions change in women as they age and experience changes in hormonal status. She felt strongly that there is immense potential to advance opportunities to work *with* our microbes instead of seeking to get rid of them: "I encourage my patients to think of their bodies like a canvas upon which the bacteria live." This is an important symbiotic relationship.

OVERDOSING WOMEN ON ANTIBIOTICS

Antibiotics are frequently prescribed for specific bacterial infections including respiratory tract infections, bronchitis, and UTIs. A 2016 analysis of eleven studies encompassing antibiotic use in over forty-four million individuals found some very surprising results. Women are 27 percent more likely to receive an antibiotic prescription in their lifetime than men. The amount of antibiotics prescribed to women aged sixteen to thirty-four is 36 percent higher than similarly aged men, and women aged thirty-five to fifty-four receive 40 percent more antibiotics than men in the same age group. Two types of antibiotics often used for respiratory tract infections, cephalosporins and macrolides, were particularly high in prescriptions to women.

We know that women have unique issues with UTIs, and at first glance one might assume that this may in part account for gendered differences. However, this doesn't hold true when examined more closely: The types of antibiotics (quinolones) usually used to treat UTIs were not increased in women compared to men. It is well documented that outpatient treatment of respiratory tract infections with antibiotics is over-prescribed, with 40 to 50 percent of patients receiving "inappropriate" prescriptions. Women generally visit their physicians more than men and are twice as likely to see their doctors for a respiratory tract infection than men, despite the fact that women do not have higher rates of these infections than men.

Antibiotics are not the only class of drugs where gendered differences are evident; medication for thyroid therapy and depression are also skewed towards women, and these differences can't be fully explained by medical reasons. Still, antibiotics show the highest disparity in prescription practices among male and female patients, so women should pay particular attention to these prescriptions. Inquire about the *medical* reasoning to make sure that the prescription is *medically* necessary. It is traditionally said that "an antibiotic might not work, but it can't hurt." Given all that we know about antibiotics and their effects on the microbiome, this concept needs serious revisiting for members of both sexes.

Looking Ahead

The realization that the bladder is not sterile is a major paradigm shift in how we think of urinary tract biology and might be leveraged into novel applications to improve bladder health, including how we treat UTIs. Older women are more likely to get UTIs, but older adults in general suffer more from bladder problems as weakening muscles make it harder to fully empty the bladder. When urine stays in the bladder longer, infections are more likely.

Vaginal health is tightly tied to the vagina's resident microbes. The ability to modify or deliver particular vaginal microbes holds much promise for better treatment of vaginal infections, in addition to enhancing menopausal vaginal health. Given that the gut microbiome can modulate estrogen levels post-menopause, in the future we will hopefully be able to improve symptoms of menopause through dietary practices, including the consumption of particular probiotics, prebiotics, and fermented foods. Because the vagina is a mucosal surface (which ties into the immune system), there is speculation that vaginal microbes might be used in vaccines, or to alter immune responses.

QUICK TIPS

- **No-douche zone:** The inside of the vagina is very capable of self-cleaning and ridding itself of unwanted fluids and bacteria. Douching disrupts the vagina's normal microbial communities and pH, which can result in irritation and lead to severe gynecological conditions.
- **Cranberry cocktail:** If suffering from recurrent UTIs, cranberry supplements may help prevent future infections because they can block the microbial pathogen's ability to adhere to host cells. Cranberry consumption represents a low risk prevention strategy—especially when compared to the prospect of taking chronic antibiotics.
- **Closing the prescription gender gap:** Women are significantly more likely than men to receive antibiotic prescriptions, and are at greater risk for sexist, gender-based medical care. Ask your doctor when she or he prescribes antibiotics; double-check that the prescription is warranted given your health condition and symptoms.

10

Microbes Meet Cancer

An editorial published in the March 2016 issue of the influential medical journal *Lancet Oncology* declared that cancer is no longer the big scary "C" it once was. The medical reality of cancer is changing as doctors and scientists come to recognize that many cancers can now be manageable for a very long time—and many more are even curable. However, the word still elicits an understandable rush of dread and visceral fear for most. A cancer diagnosis is terrifying and can exact an immense toll on the physical and mental health of patients and their loved ones. After diagnosis, life expectancy can range from months to decades.

Cancer is not a single monster of a disease. Rather, it describes more than one hundred diseases in which normal cells grow out of control. Healthy cells divide in an organized way so that when they are worn out or damaged, new cells take their place. When cancer develops, however, this orderly process breaks down. The cells divide continuously, forming the growths we know as tumours. Cancer can occur anywhere in the body—the brain, lungs, breast, colon, even in the blood. Some cancers grow and spread fast, while others grow more slowly. They also respond to treatment in different ways. The most common treatments are surgery, chemotherapy, and radiation, which can have significant side effects. But there are promising new ways to treat cancer, such as immunotherapy and precision medicine-based approaches, including sequencing the tumour, which can offer treatment with fewer side effects. From November 2015 through October 2016 alone, the FDA approved eight new cancer treatments and twelve new uses of previously approved cancer therapies, including the first liquid biopsy test, providing many new options for cancer treatment.

We are also getting better at preventing cancer through methods such as tobacco avoidance, healthy diets, regular physical activity, asbestos removal, and sun protection. Prevention is the ultimate goal, but many factors that cannot be controlled contribute to cancer. Each year, fourteen million people worldwide learn that they have cancer, and over eight million people die from the disease. Novel approaches and advances in cancer treatment are needed to reduce the global burden of cancer, and microbes appear poised to take centre-stage.

Having studied cancer for over thirty years, Dr. Shoukat Dedhar—distinguished scientist in the Genetics Unit of the British Columbia Cancer–Vancouver and professor in the Department of Biochemistry and Molecular Biology at the University of British Columbia—has seen the highs and lows of the field's evolution. When he spoke with us, he noted that scientists are beginning to realize that microbes play a much bigger role in both cancer and its treatment than anyone previously thought. Dr. Dedhar became aware of the microbiome when he learned that a microbe, not stress, contributed to stomach ulcers and cancer (a topic we explored extensively in Chapter 6). "It was surprising, and yet also unsurprising if you really think on it. We can see the influence of microbes everywhere." His list of such conditions included ulcerative colitis (a chronic inflammatory bowel disease induced by microbes that increases one's risk for bowel cancer), colon cancer, viruses involved in liver cancers, and more.

MYTH: Cancer is cellular, so microbes aren't involved.

FACT: Microbes are intimately involved in many cancers. They influence the risk for developing cancer, immune responses that affect both cancer and tumour control, disease progression, and response to treatment.

Why are we just coming to this knowledge? Evolving technologies enable researchers to better "see" the microbiome and its many interactions with cancers. Dr. Dedhar was particularly interested in observing tumour micro-environments: "Are there bacteria within growing tumours that help them flourish? What role do they play? Perhaps a tumour is more aggressive because of assisting microbes? Nothing is known." Yet that is changing fast. Emerging data suggests the possibility that colon cancers are driven by mutations in certain genes (such as ZEB2), which may promote bacterial infiltration into the tumours. Treatment with antibiotics could inhibit tumour growth. "Now *that* is an area," Dr. Dedhar said, "that I would like to explore."

Microbial Carcinogens

An estimated 20 percent of all cancers are linked to microbial agents. We saw in Chapter 6 how *Helicobacter pylori* is linked to stomach cancer; it can inject a bacterial molecule into host cells that activates a cell division pathway, which triggers the cells to divide without stopping. Gall bladder cancer is also associated with *Salmonella typhi* infections. Since the chronic inflammation triggered by microbes causes tissue damage, subsequent mutations that occur during repair can lead to cancer in both the gall bladder and stomach. In mouse models, *Helicobacter hepaticus*, a microbe that causes intestinal tissue inflammation, triggers an increase in mammary tumours in several mice strains. This is probably through activation of the body's immune system and inflammation.

As we'll see in Chapter 11, this inflammatory response is one of the key ways that microbes' impact on immune system development and function relates closely to cancers. Our immune system's defences normally detect and try to control tumours in a process known as immunosurveillance, a kind of alarm system that protects us. If the alarm system is triggered, it calls on the body's response, which in turn stops any unwanted invaders. If microbes make our immune system less efficient (think of power being intermittently cut to the alarm system), this can affect our rate of developing tumours. Another tactic our immune system uses to defend the body against invading microbes is inflammation. Activating the alarm, a pro-inflammatory response, can be pre-carcinogenic as it leads to further cell damage and faulty repair. With alarm bells increasingly ringing, an inflammatory state increases cancer risk.

So far, the International Agency for Research on Cancer has designated a total of ten microbes as carcinogenic to humans; this is actually an incredibly small fraction of the estimated trillions of microbes that are found on Earth. Most of these cancer-causing microbes are viruses, including Hepatitis C (liver cancer), human papillomavirus (cervical cancer), and Epstein-Barr Virus, which was the first virus shown to cause a wide variety of cancers.

Rapidly evolving knowledge about the microbiome's role in the body at large also suggests further microbial-cancer associations. For example, numerous risk factors for cancer—including obesity, cardiovascular disease, type 2 diabetes, and aging—have established microbial links. There is even a proposed "cancer hygiene hypothesis" based on the more general hygiene hypothesis, in which increases in certain cancers are linked to modern lifestyles that avoid microbes and encourage sterility, as well as to consumption of radiated and processed food.

We also experience risks and benefits from the many different molecules that microbes produce and secrete into the body, a subset of which can impact cancer. Several gut microbes break down fibrous food substances, for example, and produce the short-chain fatty acids (SCFA) butyrate, propionate, and acetate. These molecules can suppress inflammation, which in turn reduces cancer incidence, but other microbial metabolites can promote carcinogenesis. This includes bile acids, hydrogen sulphide gas (the rotten egg smell in intestinal gas), and modified steroid hormones. Learning more about the presence of any of these molecules in the body could be a powerful tool in our war against cancer.

COLEY'S TOXINS

The use of microbes in cancer therapy goes as far back as the late nineteenth century, when American physician William Coley began treating various tumours with a mixture of two heat-killed bacteria: *Streptococcus pyogenes* and *Serratia marcescens*. Called "Coley's Toxins," this cocktail was injected locally into the tumour. This produced significant inflammation, which could assist the body's immune system in controlling tumours. The method was applied from 1893 until 1963 in the United States, and up to 1990 in Germany. However, the results of this treatment were mixed and so treatments were stopped. Clinical trials were not conclusive, and the scientific evidence did not support it as a viable cancer treatment.

This experiment didn't die entirely, however, for the Coley's Toxins concept is behind the current treatment for bladder cancer. Bacillus Calmette-Guérin (BCG) is a strain of *Mycobacterium bovis (M. bovis)*, which is closely related to *Mycobacterium tuberculosis*. Because *M. bovis* is related but does not cause tuberculosis, it is used extensively as a live-attenuated (a fancy name for a weakened, but live microbe) vaccine for tuberculosis. *M. bovis* is the only FDA-approved treatment for bladder cancer. Live microbes are injected into the bladder following surgery. Here they presumably trigger a strong inflammatory response, which helps prevent tumours recurring. The method eradicates the cancer in 70 percent of patients.

The Fear of Lumps

Every October we see pink ribbons adorning jackets, professional athletes donning pink shoes, and buildings illuminated in pink light. The goal of Breast Cancer Awareness Month is to increase public awareness of the disease and raise funds for research, prevention, and treatment. Breast cancer affects one in eight women in the United States and is the second leading cause of cancer deaths in women after lung disease. Men are also affected: Every year just over two thousand men are diagnosed. Their mortality rate is slightly higher than in women, perhaps because awareness is lower and they are less likely to assume that a lump is breast cancer. Overall, breast cancer accounts for forty thousand deaths per year in the United States alone. Breast self-exams are recommended for both genders to check for any persistent lumps or changes in breast tissue.

Some risk factors for breast cancer can't be changed, including sex, age, and genetics. But lifestyle changes can affect other factors such as being overweight, lack of exercise, unhealthy diet, and smoking cigarettes. As with many diseases, risk of breast cancer increases with age. About two in three invasive breast cancer cases involve women aged fifty-five and older. High estrogen levels are a risk factor, especially for current or former users of HRT. This menopausal treatment typically uses a combination of estrogen and progesterone, which increases the risk of being diagnosed with breast cancer, even when used only for a short time. Estrogen-only HRT increases the risk of breast cancer, but only when used for more than ten years. Estrogen-only HRT also can increase the risk of ovarian cancer. Given these risks, HRT is generally only prescribed when women have severe menopausal symptoms that significantly reduce quality of life.

Given the link between the gut microbiome and estrogen levels, discussed in the preceding chapter, are there connections between the microbiome and breast cancer? Maybe. In a preliminary study, forty-eight post-menopausal women with breast cancer had altered gut microbiomes. They had decreased microbial diversity and altered composition compared to cancer-free controls, in addition to higher estrogen levels. There are also several epidemiological studies showing that consumption of fermented milk products (e.g., kefir, yogurt) is associated with a decrease in breast cancer risk, although it has not been established whether or not this occurs through the gut microbiome and the enzymes it encodes. We need more scientific studies to determine whether the microbiome has a direct causal effect on breast cancer risk. It may be correlated simply because decreased microbial diversity is associated with obesity, insulin resistance, and other risk factors associated with breast cancer.

Antibiotics, especially ampicillin—which is used to treat different types of bacterial infections including ear infections, bladder infections, and pneumonia—eradicate gut microbes that produce the estrogen-deconjugating enzyme activity. This means lower circulating estrogen levels and higher conjugated estrogens in the feces. There is also increasing evidence that dysbiosis associated with extensive antibiotic use (including commonly used tetracycline and sulphonamide) correlates with breast carcinomas.

Does that mean we should we avoid taking antibiotics, except when medically necessary, to protect ourselves against breast cancer? Perhaps. But there are currently no data to indicate that minimizing antibiotic exposure could directly protect against this specific disease. In a large North American study of nearly ten thousand women, increased prior antibiotic use increased the risk of breast cancer. All classes of antibiotics were associated with this increased risk. Other studies have found a slightly higher risk of breast cancer associated with antibiotic use. Antibiotic-induced changes to the microbiome may affect the metabolism of sex hormones such as estrogen, and in turn the risk of breast cancer. The general consensus is that there is a small increase in risk with antibiotics, but their use is so ubiquitous it is difficult to prove a clear link.

Alcohol consumption also increases the risk of breast cancer, especially in post-menopausal women. Again, the gut microbiome may be involved in this. There are always significant changes in the gut microbiomes of individuals with chronic alcohol abuse, including the appearance of large intestinal microbes in the small intestine, a condition called Small Intestinal Bacterial Overgrowth, or SIBO. Excessive bacteria in the small intestine is frequently implicated as the cause of chronic diarrhea and malabsorption, and SIBO patients may also suffer from weight loss, nutritional deficiencies, and osteoporosis. SIBO is also found in animal models of alcoholic liver disease. How alcohol-induced microbiome changes relate to estrogen and breast cancer is currently undefined, although several metabolites linked to estrogen production were altered in the animal models. This could influence breast cancer susceptibility if true for humans.

A surprising new finding is that there is a breast tissue microbiome with a diversity similar to that found in the gut. Researchers have observed differences in the breast microbiome of those with breast cancer compared to controls without the disease. Several small studies suggest that the breast microbiome in breast cancer patients, as well as in the control subjects, is different from nearby healthy tissues. In a small study of forty women (twenty of whom had breast cancer), particular microbiota were differentially enriched in tumour breast tissue and normal tissue. This means that scientists observed a decreased bacterial DNA load (i.e., fewer microbes) in tumour tissue, and healthy breast tissues had more microbes.

Further complicating this emerging field, another recent study—of seventy women who had breast cancer (normal adjacent tissue was also collected) and healthy controls—provided a different picture. This study found that the microbiomes from cancer biopsies and normal adjacent tissue were not different, but when compared to healthy control breast tissue, both were different. In other words, there doesn't seem to be a microbiome profile specific to tumours, but there is a different overall profile for both healthy and cancerous breast tissue from individuals with breast cancer compared to healthy controls. Women with breast cancer had elevated levels of Enterobacteriaceae (including *E. coli*) and *Staphylococcus*. Using cultured human cells in the lab, researchers also showed that these microbes increased DNA double strand breaks. This is associated with increased tumour formation, although it remains a very preliminary lab observation. Although there were no differences observed based on menopausal status, one interesting finding is that breast cancer patients' bacteria encoding beta-glucuronidase were enriched—which, as we have seen, increases estrogen levels.

To sum up, there are certainly differences in the breast microbiome of women with breast cancer compared to healthy controls, but we are not at the stage yet where we can use the microbiome as a tool to detect, prevent, or treat breast cancer. In the next few years we anticipate intensive studies in this exciting new area of breast cancer research.

Antibiotics and Cancer

Part of the argument for the "cancer hygiene hypothesis" described earlier has to do with the role of antibiotics in cancer, since several studies indicate that antibiotic-driven dysbiosis can affect cancer rates over time. For example, a cohort study in Finland observed individuals who took little to no antibiotics (zero to one prescription) in comparison to those who had six or more antibiotic prescriptions. They found that heavy antibiotic users had a 15 percent increase in risk for developing colon cancer during nine years of follow-up. Similarly, a large study of colorectal adenoma (the precursor to colorectal cancer) in female nurses found that long-term antibiotic use beginning in early to middle adulthood was associated with an increased risk of colorectal adenocarcinoma after age sixty. Interestingly, in the past four years, antibiotic use did not increase the incidence, although this makes sense given that colon cancer often takes a decade to develop.

There are also hints that antibiotic exposure may affect other cancers throughout the body. Epidemiological evidence in a large study of over six hundred thousand people indicated that antibiotic exposure may increase rates of lung, prostate, and bladder cancer. These results support exercising caution when deciding whether or not to use antibiotics. While they are very effective at treating bacterial infections, they may have longer-term consequences of increasing cancer risk. This science linking antibiotic use to cancer is still early days and we do not yet fully understand the mechanisms involved, but it certainly smacks of microbial involvement.

Colorectal Cancer

Given the many interactions between intestinal microbes and the gut, it is not surprising that there are linkages between gut microbes and colorectal cancer (CRC). Often referred to as colon cancer, CRC is the third most deadly cancer among men and women worldwide. Approximately 140,000 people are diagnosed with colon cancer and over 50,000 succumb to the disease annually. In industrialized countries, the lifetime risk of developing CRC is 5 percent, but cases are on the rise among people under the age of fifty, a group that is rarely screened for this disease. Some experts predict that colon cancer rates among people aged twenty to thirty-four will increase by 90 percent by 2030.

Approximately one-third of CRC cases are attributed to genetics, or family history. It is unclear what accounts for the remaining two-thirds of cases, but, as we will see, the long-term effects of diet and microbes contribute to the disease. Indeed, colon cancer is preventable and highly treatable—there is over a 90-percent chance of a cure when diagnosed early. CRC can take ten to forty years to develop, partly because genetic mutations in the colon accumulate over time, which can lead to abnormal and cancerous cell division. The resulting overgrowth of intestinal cells forms an adenoma: a polyp, or small growth of tissue, that is typically benign. But in some cases, cancer can start in the adenoma. Colonoscopies are used as a preventive measure to search for and remove larger and suspicious adenomas. Advanced CRC is deadly, having high mortality rates, so we emphatically stress the importance

MYTH: You do not need to be screened for colon cancer if you have regular bowel movements and feel fine.

FACT: Colon cancer is a silent killer, often without obvious symptoms. Starting at age fifty—and even earlier for those with a history of family risk—everyone is urged to have a colonoscopy to screen for colon cancer. The sooner the disease is caught, the better your chances for treatment and survival.

of being proactive about getting colonoscopies, especially after the age of fifty. Those few days of discomfort could save your life.

Besides proactive screenings for polyps, there are many risk factors for CRC you can control, and others you cannot. You won't be able to change your risk when it comes to being older, having a family history of colorectal polyps, having inflammatory bowel disease, and/or the presence of adenomas. This makes it even more critical to focus on the factors we can control, such as being overweight or obese, being physically inactive, smoking, and heavy alcohol use—all of which can be lessened with lifestyle modifications. High-fat diets and those rich in red meat and processed meats increase the risk of CRC, while diets rich in fibre are thought to be protective. This is where our microbes enter the conversation. As we learned in Chapter 8, both red meat and processed meats have proteins and other molecules that gut bacteria can modify to produce compounds that are harmful to the gut lining. They trigger mutations in cells that can lead to polyps, and ultimately CRC. Diets rich in saturated fats can also increase the body's bile acid production. While bile acid helps solubilize and break down these fats, it is also a very destructive detergent. Gut microbes strongly affect bile acid metabolism and modify bile acids to form several other compounds, which may contribute to cellular mutations that cause CRC.

Many of the immune-microbial connections discussed earlier play a particular role in the development of CRC. Dysbiosis increases the number of inflammatory microbes, which in turn triggers inflammation and subsequent CRC. In animal models, mice lacking innate immune pathways—which are key to inflammation—are more resistant to CRC. There is ongoing debate regarding whether or not particular microbes are responsible for CRC. One microbe receiving a lot of attention is *Fusobacterium nucleatum (F. nucleatum)*, a common oral microbe and minor component of the intestinal microbiota that can also trigger an inflammatory response. By studying the microbial composition of adenomas and colon tissue, scientists found higher numbers of *F. nucleatum* in colonic adenomas (polyps) and colonic tumours when compared to normal colon tissue from the same patient. When germ-free mice susceptible to tumours are inoculated with this microbe, there is an increase in tumours. *F. nucleatum* is invasive, which means it can adhere strongly to host cells and even enter them. Adding to this, the unwanted microbe has a cell surface protein (FadA) that is able to bind to a cell receptor and tell the cell to multiply uncontrollably. We need more studies to better understand these mechanisms and how to manage *F. nucleatum,* which may in fact be a "biomarker" for CRC, one we can measure to decrease the incidence of tumours or even prevent tumour progression in patients who have an increased risk of CRC.

There are other particular microbes being associated with CRC, but data in humans is not very convincing at this early stage. We need to treat discoveries critically and thoughtfully, and wait for studies that confirm findings before claiming causation. Remember, just because a particular microbe is associated with tumours (or any other disease, for that matter) does not mean it causes them—perhaps the tumour environment merely favors that particular microbe's growth. The other problem with studying CRC is that its tumours are extremely slow-growing. The microbe(s) that could have triggered it may well have long ago left the body by the time the tumour is detected. In short, it is a long road ahead for scientists to establish any causation between specific microbes and CRC incidence—but we are making steady progress.

MYTH: Everyone with cancer must start treatment immediately.

FACT: Not necessarily. Under certain conditions—if the cancer is found at an early stage, if it is growing slowly, or if treatment will cause more discomfort than the disease—doctors may recommend "active surveillance," a process where they monitor the cancer closely, and begin treatment if the cancer shows signs of growing or begins to cause symptoms. Treatment can also put patients at greater risk for infection, which is increasingly dangerous in a world of antibiotic resistance.

Cancer and Antibiotic-Resistant Infections

Infections are a major and life-threatening danger to all cancer patients given that the cancer and its therapies prevent the immune system from working at full capacity. Several pathogens that cause infections are resistant to many antibiotics, which makes treatment extremely difficult. We have used the same drugs so widely and for so long that the infectious organisms have adapted and are now often resistant to traditional antibiotics. This makes our need for new solutions dire.

Although the research is in its infancy, there is significant optimism that altering the microbiome could help counter these antibiotic-resistant infections. It is well established that a healthy microbiome can block pathogens by out-competing them for nutrients and sites to colonize. Patients undergoing any kind of treatment can work to maintain a healthy microbiome and encourage this internal environment. This includes promoting healthy, diverse microbiota by eating a diet that includes foods such as kefir and other active-culture yogurts; fermented plant-based foods (e.g., tempeh, miso, sauerkraut); fruits; and leafy greens.

A healthy microbiome also improves the gut barrier by decreasing its permeability, which further prevents unwanted microbes or their molecules from entering the body—especially from reaching the blood stream. Excitingly, patients undergoing stem cell transplants colonized with certain microbes were protected against subsequent infection with vancomycin-resistant enterococcus (VRE). There are 20,000 infections and 1,300 deaths annually from VRE. Enterococci cause a range of illnesses, including bloodstream infections, surgical site infections, and urinary tract infections. There are few or no antibiotic treatment options left, as these bacteria continue to outwit our current drug supply. This is why this discovery of microbe-boosted protection is so exciting. The potential to save lives is immense, and there is significant hope that in the future we will utilize the microbiome to decrease the risk of infections in cancer patients.

Stem Cell Transplants

Deep in the bone marrow live stem cells, which are the precursor cells forming all the cells in our blood. Sometimes, the cancer cells that arise in the bone marrow result in blood and bone marrow cancers (e.g., multiple myeloma and leukemia). To treat these cancers, all of a patient's bone marrow is destroyed through chemotherapy or radiation. Blood-derived stem cells from a related donor, such as a sibling, are then placed into the patient where the stem cells localize to the bone marrow and "reboot" the production of normal blood and immune cells. This technique is called Allogeneic Hematopoietic Stem Cell Transplantation (Allo HCT). It is still fraught with problems such as infections and graft-versus-host disease (GVHD), where the donor's immune cells mistakenly attack the recipient's body. The condition can range from mild to life-threatening. And the older the person, the higher the risk for GVHD.

Several recent reports suggest that the microbiota may influence the outcome of bone marrow transplants before *and* after the procedure. Pre-transplant dysbiosis has been linked to higher risk of infections, GVHD, and reduced overall survival. In a study of eighty recipients, low bacterial diversity was also associated with significantly worse outcomes following transplantation. This suggests that a robust microbiome is important to help prevent these complications. In an exciting recent clinical pilot study, fecal transfers were successfully used to treat GVHD following transplantation. This represents a major and unexpected step forward for the field, and a groundbreaking avenue towards more effective cancer treatments.

Increased GVHD mortality is also associated with the broad spectrum antibiotic use that often follows bone marrow transplantation and would also decrease microbial diversity. Because patients' immune systems are destroyed by the procedure, they are at high risk of infection and need to rely on strong antibiotics to fight off infection. Breaking this cycle of antibiotic dependence is crucial, and microbes can provide the much-needed new methods to boost the immune system first and lessen problematic dependence on antibiotics after. In a retrospective study of 541 microbiomes of Allo-HCT transplantation individuals, a particular microbe, *Eubacterium limosum*, was associated with a decreased risk of relapse and progression of the disease. Interestingly, this microbe is also found extensively in centenarians.

There is exciting potential in applying this knowledge to medical practices. We may be able to better predict clinical outcomes based on individual patients' microbiome composition. We could also replenish the microbiome post-transplantation to reduce intestinal inflammation and improve health outcomes.

Cancer Therapy and Microbes

Besides surgical removal, traditional cancer treatments include radiation therapy and chemotherapy, which use toxic chemicals and work by preferentially killing fast-growing cancer cells. Given the toxicity of these treatments, it is not surprising they both significantly diminish microbial diversity.

One promising potential new solution to this problem is using our microbiota to predict the success of chemotherapy. Microbes can affect the bioavailability of chemotherapy agents by affecting either drug metabolism or uptake of drug. In animal studies using chemotherapy, subcutaneous (under the skin) tumours failed to respond to chemotherapy following antibiotic treatment. Cyclophosphamide, a common chemotherapy agent, required a healthy gut microbiota to achieve efficacy. Studies indicated that the microbiota were needed to potentiate the full activity of the immune system. A recurrent theme is that a strong immune system is essential to control tumour growth, and microbes can influence this immune response. We are awaiting studies in humans to confirm these findings.

What has most excited the entire oncology world is the potential of the microbiome not just to predict but also to *improve* immunotherapy outcomes. To explain this, we need to take a couple of steps back. Cancer immunotherapy works by blocking checkpoint inhibitors—pathways that impede the immune system. It essentially boosts the body's natural defences to fight cancer by unleashing the immune system. Immunotherapy improves or restores immune system function to stop or slow the growth of cancer cells, and to stop cancer from spreading to other parts of the body. There are several types of therapies, including monoclonal antibodies, nonspecific immunotherapies, oncolytic virus therapy, T cell therapy, and cancer vaccines. Immune therapies are relatively new and are often used after traditional treatments of radiation or chemotherapy have failed. The treatments work in some cases, but not in others.

Two papers published in late 2015 upended the entire immunotherapy field and brought the role of microbes in traditional cancer treatments to the forefront by showing that microbes can affect cancer therapy outcomes. In the first paper, Dr. Thomas Gajewski led a team of researchers at the University of Chicago. They asked the fairly simple question: Will manipulating microbial composition change the efficacy of an immunotherapy? His team tested this question on an immunotherapy that uses an anti-PDL1 monoclonal antibody as a checkpoint inhibitor. PDL1 constrains the production of specific immune cells (CD8+ T cells) that actively seek out and destroy tumours. Dr. Gajewski's team began by looking at the anti-tumour effect in a strain of lab mice that were bred at two different mouse facilities. These mice purposefully had different gut microbiota. The scientists then implanted melanoma tumours in the mice. They found a difference: Tumours grew less aggressively in mice from one breeder than the other. They also found a more robust T cell response in the mice with slower tumour growth. These experiments suggested that different microbiota affected T cell responses differently, and, importantly, affected tumour growth rate. To confirm this result, they let the mice live together, which enabled the mixing of the microbiota, due to general contact and the fact that mice are coprophagic, meaning that they eat each other's feces—the ultimate FMT! After swapping microbes, the two mice strains no longer had detectable tumour growth differences. Similarly, when feces were deliberately transplanted from one mouse strain to another, the anti-tumour and T cell effects were also transferred. By combining a fecal transfer and immunotherapy into more susceptible animals, even greater tumour control was observed. This suggested that the microbiome can play a central role in cancer immunotherapy treatment.

This discovery sparked another question: Are any particular microbes specifically involved in this effect on the immunotherapy? By sequencing the gut microbiota, these researchers found that *Bifidobacterium* species were the ones linked to anti-tumour immune responses. When they added a cocktail of these microbes to the susceptible animals, they were able to transfer the ability to control tumours in susceptible mice to the same extent as a fecal transfer. What really cinched things was that when they fed the *Bifidobacterium* mixture to mutant mice that lacked CD8+ T cells, they were unable to control tumours—meaning that both CD8+ T cells and this microbe are needed. Further, if they killed the bacteria before giving it to mice, there was no effect. This suggests that live bacteria are needed for the effect to work.

The second study, led by Dr. Laurence Zitvogel in France, offers further evidence—but with different players. His team used lab mice with various sarcoma, melanoma, or colorectal tumours to investigate the effect of immunotherapy against a different checkpoint inhibitor, CTLA-4. This immunotherapy is approved for treatment of patients with metastatic melanoma (skin cancer). Dr. Zitvogel's team found that neither germ-free nor antibiotic-treated animals responded to the anti-CTLA-4 therapy. The normal mice (with normal microbiota) responded well. Researchers also found that if they added *Bacteroides fragilis* (*B. fragilis*) to germ-free or antibiotic-treated animals, the effectiveness of the immunotherapy was restored; this effect was mediated by T cells. In another analysis, studying the gut microbiome of twenty-five skin cancer patients, researchers found that a fecal transfer into germ-free mice of patients' feces that contained *B. fragilis* resulted in a restoration of the anti-CTLA-4 anti-tumour activity.

Two more papers were published back-to-back in late 2017 in *Science* that validated the concepts seen in animal models in humans. Extending their mice findings, Dr. Zitvogel's group showed that resistance to immune checkpoint inhibitors is due to the microbiome, and that antibiotics blocked the efficacy of PD1 inhibitors. They also found that they could transfer feces from cancer patients, who did or didn't respond, into mice, and show that response to checkpoint inhibitors (or not) could be transferred with the feces. The team further analyzed the microbiome content of cancer patients and found a beneficial correlation to *Akkermansia muciniphila*. They could even spike non-responder feces with this microbe, and in mice, they would now respond to PD1 immune therapy. In the other paper by Dr. Jennifer Wargo's group, the gut microbiomes of 112 melanoma cancer patients were analyzed. Researchers found that the microbiome of responders to anti-PD1 immunotherapy had a much higher diversity, and an enrichment in Ruminococcaceae organisms. They also showed that the positive response to checkpoint inhibitors could be transferred into mice simply by a fecal transfer.

Collectively, these findings have unleashed a flurry of research and excited speculation for cancer therapy. Many scientific groups are now exploring these concepts and additional clinical trials are underway. While we need a better understanding of exactly which microbes are involved in cancer and how, many believe that very soon patients being considered for immunotherapy will undergo a microbiota analysis first, and those lacking particular microbes may need to be given certain microbes to enhance treatment. This represents a huge step forward in the pursuit of more effectively preventing and treating cancer.

QUICK TIPS

- **Boost your defences:** A healthy microbiome can help block pathogens and potentially improve cancer treatment outcomes. Maintain a healthy microbiota by eating a range of fresh, whole, and fermented foods.

Consider taking a probiotic supplement—especially if you are currently undergoing any treatments.

- **Get screened:** Colon cancer is a preventable and highly treatable cancer when caught in the early stages. Starting at age fifty, everyone should have regular colonoscopies to screen for colon cancer. Begin screenings even earlier if there is a history of family risk.
- **Stand up to superbugs:** Given that heavy antibiotic usage is linked to greater risk of cancer and infections, take antibiotics only when necessary and exactly as prescribed: Do not skip doses, and complete the course of treatment even if you start to feel better—otherwise you increase the chances of increased antimicrobial resistance to the antibiotic, requiring additional treatment. Do not share or use leftover medications. Shore up your immune system after a course of antibiotics with probiotic and prebiotic foods.

11

Microbe Tug-of-War: The Immune System

Every second of your life, you are under attack. Trillions of bacteria, viruses, and fungi are trying to make your body their home. In response, we have developed a complex army with sentinels, guards, soldiers, intelligence, weapons, factories, and communications, called the immune system. A finely tuned fighting machine, it protects us from the dangerous world of deadly microbial pathogens, scours our body for tumours, and remembers previous dangers—from even decades ago. It is our personal bodyguard: Ready to protect us from infection, and pounce again if any dangers reappear.

So, how does it work? Let's say you're out for a hike on a beautiful summer day, and you trip on a wooden bridge crossing a stream and scrape your leg on a rusty nail. The immune system's first line of defence—your skin—has been penetrated. Nearby bacteria seize the opportunity to enter through your open wound. They multiply inside your body's warm, moist environment, at first flying low under the radar. No alarm bells go off. But then the bacteria become so numerous they start to attack your body. The immune system kicks in, trying to stop them as quickly as possible.

Your sentinel guard cells, known as macrophages, are the first to intervene. These cells patrol and protect every crevice of the body. Most of the time the macrophages alone can stop an attack by devouring and breaking down (i.e., killing) the intruders. On top of that, they muster help from cells in the blood vessels, summoning their release and other infection-fighting molecules into the battlefield; this process is what we see and feel as inflammation in the form of redness, swelling, warmth, pain, and—in certain cases—fever. An entire immunity army coordinates to attack foreign objects, stabilize your body, and prevent illness. It is a smart, efficient, and wonderfully complex system. This doesn't only happen when we trip and fall, however. The immune system is constantly in tune with the environment regardless of acute injury: It adjusts inflammation levels up and down depending on how much of a threat is encountered and how successfully it can control the invaders.

Because microbes are such a key part of environmental exposure, they are intimately intertwined with it and its function. The first interactions between microbes and the immune system occur immediately—when you are born. Maternal vaginal and fecal microbes encountered during birth jumpstart the immune system, and in our first few months contact with our mother's microbes shape our immune system, ultimately determining whether allergies, asthma, and eczema may arise years later. Microbes and the immune system are in a constant tug-of-war throughout our lives. How do we prevent microbial pathogens from entering the body, yet tolerate the trillions of harmless and often beneficial gut, mouth, and skin microbes? In this chapter we will explore how the body achieves that balance, as well as the startling effects that microbes can have on the immune system to both our benefit and detriment.

IMMUNOLOGY CHEAT SHEET

Discussing the immune system involves many components. Here are the most frequent terms we use throughout the chapter:

- **ANTIBODY:** These protective proteins, also called immunoglobulins, are produced by the immune system in response to the presence of a foreign substance such as a pathogenic microbe (an antigen). Antibodies latch onto antigens to remove them from the body.

- **ANTIGEN:** Any substance foreign to the body that stimulates an immune response, ranging from pollen to infectious microbes. Antigens are targeted by antibodies.
- **B CELLS:** Also known as B lymphocytes, these cells are a type of white blood cell. They help the adaptive immune system by binding to a specific antigen, against which it will initiate an antibody response.
- **CYTOKINE:** These small proteins are very important to cell signalling during inflammation. They aid cell-to-cell communication in immune responses and stimulate the movement of cells towards sites of inflammation, infection, and trauma.
- **IgA:** Immunoglobulin A is an antibody that plays a critical role in the immune function of mucous membranes. These specialized antibodies selectively target and kill invading microbes.
- **IMMUNE CELL:** The overarching category of cells that make up the immune system, including types of white blood cells—lymphocytes (T cells, B cells), neutrophils (one of the first cell types to travel to the site of an infection), and monocytes/macrophages.
- **LYMPHOCYTE:** A type of white blood cell that responds to foreign invaders in the body; some lymphocytes work alone, while others coordinate with other cells.
- **MACROPHAGE:** A type of large immune cell whose job it is to engulf and destroy any invading microbes. They are some of the first immune cells to be recruited to the site of infection.
- **PATHOGEN:** Any biological agent that causes disease or illness to its host, such as a bacterium, virus, or fungus.

- **T CELLS:** Also known as T lymphocytes, they play a central role in determining the specific immune response to antigens. Types include:
    - **CYTOTOXIC T CELLS:** Destroy virus-infected cells and tumour cells.
    - **REGULATORY T CELLS (Treg):** Act to control immune reactions.
    - **T HELPER CELLS:** Assist other white blood cells in immune system processes. This includes T helper 17 cells (Th17), a subset of pro-inflammatory T helper cells.

Microbes and the Immune System

Since microbes are part of what triggers our immune systems, would we be able to live without microbes in a blissful, pathogen-free existence? Although possible (with some caveats), it is certainly not ideal, as was seen in an experiment undergone by David Vetter, who was born in Texas in 1971. Prior to birth, David was diagnosed with severe combined immunodeficiency (SCID), a hereditary disease that severely compromises the immune system. Patients with SCID die very young because they are abnormally susceptible to infections, and exposure to typically harmless microbes can be fatal.

Vetter's parents and doctors unfortunately knew what to expect because David's older brother had been born with SCID but died at seven months old. The only way for David to survive was to keep him in a sterile environment from birth and hope that a donor could be found for a bone marrow transplant to reconstitute his immune system (many of our immune cells originate in bone marrow). David was born by C-section to keep him sterile and immediately placed in a sterilized cocoon bed designed especially for him. A plastic germ-free environment would be his home for twelve years.

David occasionally ventured out into the world in a transport chamber, seven times to be exact, wearing a NASA-provided space suit. Eventually he received a bone marrow transplant from his younger sister, who was born without SCID, but he died a few months later from an undetected virus in the sister's bone marrow. David's story emphasizes just how critical an immune system is for survival in this world.

Although David was the only human on whom this experiment was performed, scientists have conducted similar exercises in animals since the 1950s. These required the invention of germ-free mice, who remain sterile from birth. Through various experiments using both these germ-free mice and antibiotic-treated mice, we realized that microbes are critical for normal development of the immune system. The experiments revealed a specialized immune tissue in the intestine called the gut-associated lymphoid tissue (GALT). GALT is the immune system's first line of defence in the gut. In germ-free animals, the GALT part of the immune system is poorly developed. The animals have fewer helpful immune cells in the intestinal wall. Similarly, mesenteric lymph nodes (where immune cells drain to) are also smaller and fewer in number. Even intestinal epithelial cells—the cells that make up the intestinal wall and form the intestinal barrier—have fewer receptors that recognize microbial products. Germ-free animals have fewer antibodies and immune cells. What all this information suggests is that microbiota are critical for normal immune system development and function.

We are only beginning to understand which microbes are needed for particular immune system functions. In many cases it may not be a specific microbe, but rather the presence of typical common surface molecules (especially LPS and peptidoglycan) that trigger immune system development early in life. In a particularly ambitious and expensive experiment, investigators colonized individual germ-free mice with fifty-three different microbiota species normally found in the human intestine. They found that each of the different species had different effects on immune system development and activity. To contemplate the various permutations and combinations found in a normal microbiome (which contains several hundred different types of species), sorting out individual effects is extremely difficult. However, as we will see, particular microbes have now been identified that affect key cells of the immune system including regulatory T cells (Tregs) and Th17 T cells. Clues about the critical role microbes play in controlling the immune system should allow us to better address numerous immune diseases that have become so prevalent in the world today, including allergies, asthma, autoimmune diseases, and inflammatory disorders.

MYTH: Airplane air makes you sick.

FACT: The recycled cabin air on airplanes is filtered by high-quality filters and is not the direct source of you catching a cold. What could make you sick is being packed on a plane with other people—who have a cold or other viral infection—coughing, sneezing, or even talking nearby. Those up to two to three feet away could make you sick if droplets containing a virus become airborne. Fatigue, which many of us experience while travelling, may also dampen your immune system: Think sleep deprivation from jet lag, crack-of-dawn early flights, and the elusiveness of sleep when cramped into a tiny transoceanic seat. When travelling, wash your hands frequently with soap and water and prioritize sleep to help out your immune system.

Microbial Conversations with the Immune System

The intestine was historically understood as a major part of our immune system's defence system against invading pathogens. It acts as a major physical barrier inside our bodies, offering multiple ways to attack microbes before they can cause an infection, similar to a "demilitarized zone" with extremely high security surrounding a nearly insurmountable barrier. Should a microbe attempt to penetrate this zone, it is met with an arsenal of antimicrobial responses from the immune system. First, the intestine has a nearly impenetrable mucus coating. Second, it secretes antimicrobial peptides and specialized antibodies called IgA that selectively target and kill invading microbes. Third, immune cells such as macrophages patrol the gut, searching for hidden microbes to destroy. Fourth, the intestinal cells have microbial sensors that immediately recognize microbial signature molecules to trigger inflammatory responses.

As we learn more about the gut immune system, our concept of how the immune system works in the gut has changed. Given that the immune system evolved in a sea of microbes, animals had to figure out ways to differentiate between ubiquitous harmless and potentially beneficial microbes and the few that could kill them. We now realize that there is a constant but small microbial seepage across the gut barrier. Previously we thought that this occurred only during infections. The immune system is in constant surveillance mode. It does not turn on and off—instead, it is always idling, sometimes revving up as needed. That way it can immediately put its full force behind eradicating serious microbial threats should they arise.

A major question still dogging scientists is how the microbes actually "talk" to the immune system. Although there are some examples of certain microbes having close contact with epithelial and/or immune cells, the method of choice is for bacteria to make molecules that can diffuse to receptors on host cells that recognize these signals and trigger defined effects. By far the best-studied class of molecules are the short-chain fatty acids (SCFAs), which we discussed in significant detail in Chapter 7. They are mainly acetate (a small two-carbon molecule), propionate (three-carbon chain), and butyrate (four-carbon chain). These molecules are produced by certain members of the microbiota that break down dietary fibre and produce SCFA. As discussed previously, SCFAs have a variety of functions. First and foremost, they are absorbed by intestinal cells and used for energy. Second, they are "anti-inflammatories" that dial back excessive inflammation, which they achieve by affecting the size and function of the network of regulatory T cells (Tregs)—the specialized immune cells that balance the immune response. This in turn dampens effector T cells that generate inflammation. Third, SCFAs communicate with various cells in the body to improve tissue resiliency. We now realize that microbiome-derived SCFA are central to most aspects of the immune system and its responses, linking microbes to the immune system via these potent molecules they produce.

On the flip side, we also now realize that the immune system has some control over shaping microbial composition. Under abnormal conditions such as an inflamed gut (e.g., IBD), where there is excessive inflammation in the intestine, or in an immunocompromised host lacking a normal immune system, there are major effects on the microbiota composition in the gut. How exactly the immune system achieves this is not well understood. Some, but not all, microbes are tagged with IgA antibodies, which would hasten their clearance by the immune system, while antimicrobial peptides that kill bacteria target certain groups of microbes, which could also influence the microbial composition.

Microbes and Advanced Immunity

Immunologists were among the first non-microbiologists to embrace the microbiome and realize that it plays a central role in normal body function. This started with germ-free mice, which as we saw above, had poorly developed immune systems. Soon, experiments started to show surprising results; using antibiotics, fecal transfers, or cohousing mice with mice that had different microbes changed the animals' microbial composition, yielding remarkable immune effects. Nowadays, most immunology meetings contain significant microbiology components as the two fields tightly intersect, providing much excitement and a new way to look at the immune system and its functions.

Typically, immunologists concern themselves with the two main parts of the immune system: Innate and acquired immunity. Innate immunity refers to nonspecific defence mechanisms that are ready to act immediately or within hours of an invading microbe's appearance in the body. These defences include physical barriers such as the skin, antimicrobial chemicals in the blood, and immune system cells such as macrophages that attach to foreign microbes. This system is found in nearly all multicellular organisms, from flies to humans. Innate immune responses are activated by the conserved microbial signatures of invading microbes, including the potent activators LPS and peptidoglycan.

Because the innate immune system works to defend all animal and plant life, its response to invading microbes is more generic. In the gut, specialized intestinal epithelial cells called goblet cells secrete copious amounts of glycoproteins (proteins that have large sugars attached to them) to form the all-important protective mucus layer. In the colon, we have two layers of mucus: The inner layer forms a near-impenetrable mucus barrier, which keeps microbes separated from the intestinal barrier and contains relatively few microbes. The outer layer is looser and contains several microbes that happily chew on mucus as a food source. This provides both the microbes and their host (us) with energy. In the small intestine, where there are fewer microbes, the mucus layer lacks clear inner and outer layers. Embedded within the mucus layer are antimicrobial peptides produced by specialized underlying epithelial cells called Paneth cells. These tiny molecules kill bacteria in general, or major subsets of them, much like a broad-spectrum antibiotic. The underlying intestinal epithelial cells also have a set of molecules that are designed to detect microbial signatures in case microbes penetrate this barrier. If activated, they send out messages (cytokines) that trigger inflammatory responses to further help control invading microbes.

In most cases, the innate immune system does most of the hard work to keep us alive. But if this line of defence is insufficient or should fail, acquired (adaptive) immunity comes in with reinforcements. Because of its complexity, only vertebrates (animals higher on the evolutionary ladder) have this added level of protection—simpler organisms have to rely on their innate immune systems alone to survive. Acquired immunity recognizes *specific* molecules on *specific* microbes using antibodies (made by B cells) and particular T cells that target certain microbes. Although it takes at least a week for this system to boot up (we rely on the innate system to keep us alive that long), the adaptive immune system can create a long-term strategy for the army of immune cells and antibodies to specifically seek out and destroy incoming pathogenic microbes.

This slick system also has the ability to remember previous encounters with a specific pathogen. After the first encounter, our body does not need to design a new defence, but can refer to the previous all-out assault tactic. It happens because of the B cells, whose antibodies recognize particular molecules on microbes and neutralize the microbe expressing it. T cells then orchestrate the immune response by secreting cytokines to modulate inflammation. Subsets of T cells also directly kill host cells that harbour viruses and other intracellular microbes that are inaccessible to antibodies (called cytotoxic T cells). Having this additional layer of immune complexity allows jawed vertebrates like humans and other mammals, birds, and reptiles much more flexibility when dealing with pathogens, vaccines, etc. Acquired immunity also gives us a "backup" in case the innate system can't handle it.

VACCINES

No chapter on the immune system is complete without some discussion of vaccines, which exploit the immune system to give us further protection against some of the deadliest infectious diseases. As we saw in Chapter 5, it is highly recommended that you get your yearly influenza vaccine. Yet reports indicate that two-thirds of Canadians do not get the flu vaccine each year, and just 59 percent of children and 43 percent of adults in the United States get theirs. In mouse models, regular flu vaccines decrease inflammation, which contributes to inflammaging, cardiovascular disease, and dementia. Mice that were vaccinated against the flu early in life were less likely to get chronic diseases as they aged. Unvaccinated mice got chronic diseases earlier, and when they got the flu, it accelerated the appearance of chronic diseases.

Although there are many reviews and commentaries in the literature about the potential role of the microbiota in vaccine responses, there are very few actual studies. A study of macaques showed that monkeys who responded better to vaccines had a stable and more diverse gut microbiota. A small human study showed that vaccine responses were better in those with greater community richness and greater diversity of their gut microbes. Given the ability of microbes to tweak the immune system, it is very probable that the microbiota affect vaccine responses. There is significant hope that we might harness microbes to further increase positive vaccine responses.

Discovery Mode

In the past decade, we have come to realize that microbes have a profound effect on shaping the acquired immune system. Even more exciting: There are now at least three examples that identify particular microbes or their products as critical to shaping the immune system. This suggests that in the future we can use this knowledge to attempt to harness and shape immune responses.

The first major leap in our knowledge of microbes influencing T cells came in 2005, when Drs. Sarkis Mazmanian and Dennis Kasper showed that a common gut microbe, *Bacteroides fragilis* (*B. fragilis*), could shape the ratio of T cells. *B. fragilis* affected the ratio of how the immune system is balanced (called the Th1/Th2 balance), which is critical for a normal immune response. Germ-free animals make increased Th2 cells (which increase allergic responses), and by colonizing these animals with *B. fragilis,* the T cell ratio became normal. Even more surprising was that one of the microbe's surface molecules, a capsular polysaccharide called Polysaccharide A (PSA)—which is a sugar on the outside of the bacterium—could repair T cell balance all by itself. *B. fragilis* that lack PSA colonized mice but were unable to restore this T cell defect, suggesting this molecule alone could correctly balance the immune system. We now know that PSA also triggers the production of

IL-10, which is an important anti-inflammatory cytokine that dampens excess inflammation. For example, PSA can protect against experimental colitis, a disease with similarities to IBD, by decreasing gut inflammation via the T cell response. These findings were a major paradigm shift in the field of immunology as they identified a particular microbe, and a molecule it makes, that affects T cell functions. It was the first link between microbes and the acquired immune system.

Around 2006, there was another leap forward in this field, which we discussed with Dr. Dan Littman, professor of molecular immunology in the Department of Pathology and Department of Microbiology at New York University. It came about when a postdoctoral fellow in Dr. Littman's lab, Ivaylo Ivanov, was looking into recently discovered Th17 cells. These cells were found to be critical in promoting inflammation, particularly in autoimmune diseases. They play important roles in the immune response and comprise 30 to 40 percent of the T cells in the gut. Germ-free animals lacked Th17 cells (three groups simultaneously showed this), but nobody knew what might trigger their formation.

Ivanov had a hunch that microbes might influence Th17 cells since these cells were particularly abundant in the intestine. To investigate his hypothesis, he treated his mice with various antibiotics and successfully demonstrated that antibiotics could influence Th17 cell levels. Different antibiotics had different effects, hinting that a particular microbe or group of microbes could be involved.

Dr. Littman recounted that Ivanov then made yet another surprising and completely accidental discovery: "He noticed that genetically identical mice from two different standard mouse suppliers had very different levels of these cells. The differences were five to tenfold!" Ivanov even showed that if mice from one supplier were cohoused with mice from another, the levels of Th17 cells changed (remember mice are coprophagic and are swapping microbes through eating each other's feces). This finding had profound implications for immunologists. Until then, it had been common practice for scientists to order a special gene-mutated mouse (a "knock-out mouse") from one supplier, and the complementary strain with no mutation (the "parental" strain) from another supplier, assuming they would be identical except for the mouse gene mutation. Or scientists could even compare newly arrived special mice to regular ones that had been in the lab for years. Ivanov's hunch forced immunologists to cohouse mice in cages to normalize their microbiota, and to recognize that the microbiome has profound influences on the immune system. It has also called into question fifty years of immunology results based on mouse immunology.

Once Littman's group had shown that microbes are involved in the production of Th17 cells, the quest began to identify which one(s) were responsible. However, this was easier said than done. In those days, the tools to identify particular microbes were in their infancy and not wholly reliable. In an effort to address this, Brett's lab at the time used particular microbe DNA sequences coupled to fluorescent probes to make microbes visible under a microscope, thus enabling the researchers to get a general idea of the types of microbes that were present. This technique may have been crude, but it provided an indication of the general microbial composition.

Brett will never forget the day when Dr. Littman excitedly called him up and asked for help identifying the guilty microbe. Littman's lab had stained the microbes. In the animals that had Th17 cells, they could see long, skinny, segmented microbes that were lacking in those animals missing Th17 cells. Way back in 2000, the Finlay lab had studied rabbits that carried these segmented filamentous bacteria (SFBs) and were more resistant to pathogenic *E. coli* infections, but had no way of studying them other than looking at them under powerful microscopes. So now the scientists rolled up their collective lab coat sleeves and showed that, yes, these SFBs seemed to be associated with Th17 cell production. The microbes were different between Th17-producing and Th17-lacking animals.

A year later, Ivanov and Littman, along with Dr. Kenya Honda, proved that SFBs were responsible for Th17 cell production, and that they were buried in the mucus in intimate contact with underlying epithelial cells in the small intestine. These microbes are now grown in labs around the world, and intensive studies are underway to determine exactly how they shape the immune system.

The third chapter in the microbiome/T cell story came out of Dr. Honda's lab in Japan in 2011. Using similar techniques with germ-free mice, the team showed that a collection of microbes (forty-six in all) belonging to the genus *Clostridiales* were needed for the proper balance of Treg cells. Even more importantly, they showed that seventeen specific human isolates (but again not a single strain) of *Clostridia* had the same effect in mice. This group of human microbes is now being commercialized as a potential way to modify the immune system to treat various diseases such as IBD and other immune diseases.

Collectively, these three breakthroughs (and many more) demonstrate that microbes heavily influence T cell development and function (one of the two main arms of the acquired immune system). Dr. Littman recounted that it is an exciting time as we begin to understand how different bacteria generate different types of immune responses. We are beginning to turn to animal models of autoimmune diseases to see that certain bacteria can amplify—and even cause—symptoms of the disease. In a study on rheumatoid arthritis, for example, mice did not get sick if they were treated with antibiotics or were depleted of all the microbiota (germ-free). The take-home message is that contact between bacteria and the immune system in the intestine can lead to systemic (whole-body) reactions via the immune system. This in turn can contribute to autoimmune diseases in sites very distant from the gut.

In addition to T cells, the other major component of the acquired immune system, B cells and the antibodies they produce, is also affected by microbes. Germ-free animals are defective for IgA antibody production. IgA is an antibody type that is secreted into the gut and has been shown to influence the microbiome composition by recognizing specific subsets of microbes. If a host makes IgA to a particular microbe, it means that the microbe has breached the mucus barrier and triggered an immune response—meaning the immune system has been alerted to its presence, which also indicates it is in intimate contact with the body, not just hanging around the lumen of the gut. This is a way of stopping invading microbes from entering the body, and also a useful way of identifying microbes that directly contact the immune system.

When germ-free animals are transplanted with microbes, IgA production increases markedly. On the other hand, we see that IgA-deficient people naturally have more "inflammatory" microbes as part of their microbiota, and IgA-coated microbes transferred into animals seem to trigger inflammatory responses. We hope that such techniques can be used to further hone in on the inflammatory microbes in diseases such as IBD, as the IgA gives clues as to which microbes the body is responding to.

The current paradigm shift in the field of immunology makes for exciting times in microbiology. As Dr. Littman put it: "What is now clear is that there are particular microbes we evolved with that are necessary for everyday functioning. Our microbes can lead to very beneficial results given their interconnections with the immune system. I think we can harness this knowledge to treat autoimmune and other diseases." While we still do not know all the mechanisms at work, Dr. Littman eats a yogurt every morning with his microbes in mind: "Hopefully we'll be able to make better yogurt soon to keep us young for a lot longer!"

Autoimmune Diseases

The immune system is designed to detect and destroy foreign microbes and particles, and as we've seen throughout this chapter, for the most part it does a terrific job. However, to function properly it has to be able to tell the difference between friend and foe: A foreign organism versus one doing no harm happily hanging out in our body. In the case of autoimmune disorders, the immune system mistakes internal body molecules and tissue for intruders and then organizes an attack on these supposed external threats. This causes the body to attack itself for no apparent reason, the phenomenon behind a series of often-serious diseases called autoimmune diseases. They are quite varied in nature, and we will go through some of the major conditions below: rheumatoid arthritis, ankylosing spondylitis, lupus, multiple sclerosis, and gout. Now that we know microbes' intimate connection to immunity, we have increasing evidence of how the microbiota plays a role in many autoimmune diseases. This might lead to promising new treatments for these patients.

Rheumatoid arthritis

Rheumatoid arthritis is a chronic inflammatory disorder that affects the joints, including the hands and feet. The body makes autoantibodies that attack joint components; this leads to inflammation in synovial joints (the lubricated movable joints, such as elbows, wrists and ankles), deforming damage to cartilage and bone, and eventual disability and increased mortality. Age is a major risk factor for this disease—it usually starts in forty to sixty-year-olds—but there are also genetic and environmental factors that we do not yet fully understand. We think that in the disease's development, the equilibrium between pro- and anti-inflammatory pathways may become unbalanced. An increase in Th17 cells, coupled with a decrease in Treg cells, would result in increased B cell production. This in turn leads to increased autoantibodies produced by the B cells, which seems to lead to rheumatoid arthritis.

Several lines of evidence point to microbial involvement in rheumatoid arthritis. In several animal models, germ-free mice remain healthy, while the introduction of gut bacteria trigger the disease. Antibiotic treatment in animal models alters the outcome of disease, either worsening or improving it, depending on the antibiotic. Given that segmented filamentous bacteria (SFB) trigger the production of Th17 cells (as discussed earlier in this chapter), putting SFB into germ-free animals triggers the disease in a mouse arthritis model. There is also a strong correlation between periodontal disease and rheumatoid arthritis: The dental pathogen *Porphyromonas gingivalis* (discussed in Chapter 4) is now recognized as a link between periodontitis and joint inflammation.

Several studies have shown that patients with this disease have dysbiosis of both their oral and gut microbiota. In a study of 114 individuals, the presence of *Prevotella copri* strongly correlated with disease. The awareness of microbial involvement has increased hope that additional therapies can be designed to help treat and control rheumatoid arthritis. However, like most things microbiota-related, we await more studies and information before we can put new therapies into practice.

Ankylosing spondylitis

Ankylosing spondylitis is another autoimmune arthritis and is characterized by long-term inflammation of the spine. It eventually results in a curved spine and stooping posture. Over 90 percent of those with the disease are positive for HLA-B27 (a human gene marker found in only a fraction of the population), although there are hints that environmental/microbial factors may also be involved in this disease. Patients with this disease have documented dysbiotic gut microbiota. They also have increased gut permeability and local and systemic inflammation. Inflammatory microbes such as *E. coli* and *Prevotella* are increased in the ileum area of the small intestine in patients, which presumably promotes further inflammation. Similar to inflammaging, scientists propose that dysbiosis increases gut permeability, which then allows inflammatory bacterial products to enter the bloodstream. This, along with altered host genetics, triggers the inflammation associated with ankylosing spondylitis.

Lupus

Lupus (systemic lupus erythematosus) is an autoimmune disease caused when the immune system attacks its own tissues in the body. It can affect the joints, skin, kidneys, blood cells, brain, heart, and lungs. It is more prevalent in women: Studies show a range from 1.2:1 to 7:1 odds that women are more likely to have lupus than men. A red face rash, often one of the symptoms, gave the disease its name. *Lupus* means wolf in Latin, and the face rash apparently resembles a wolf bite. People with this disease will have periodic flare-ups of such symptoms, including fatigue, joint pain, rash, and fever. Like all the autoimmune diseases we discuss, there is a strong genetic factor to this disease, in addition to an environmental—microbial—one. In a study of sixty-seven lupus patients and sixteen healthy controls, lupus patient gut microbiota was less diverse, had an increase in Proteobacter (inflammatory microbes), and a decrease in *Firmicutes* (generally anti-inflammatory). Interestingly, patients also had an increase in *Prevotella copri*, which also had a strong signal in rheumatoid arthritis patients (see page 210). Also similar to rheumatoid arthritis was an increase in Th17 cells, the primary immune cell responsible for the inflammatory responses and tissue destruction seen in lupus. These are clues that correcting and restoring the gut microbes to improve T cells (increased Tregs and decreased Th17 cells) may provide new tools to help treat this disease. Future microbial therapies may be applied to reshape the immune system of lupus patients.

Multiple sclerosis

Multiple sclerosis is a disease in which the immune system eats away at the myelin sheath, a protective covering of the nerves. The resulting nerve damage disrupts communication between the brain and the body. Treatments such as physical therapy and medications can suppress the immune system to help with symptoms and slow disease progression, but it cannot be cured. Symptoms can include vision loss, pain, fatigue, and impaired coordination.

Animal models have shown that changes in T cells—including increases in Th17 cells and decreases in Treg cells—result in changes in autoimmune inflammation, including autoantibody production by B cells. Germ-free mice are markedly less at risk for the disease. However, if SFB are added to germ-free mice, Th17 cells increase—and this is sufficient to trigger disease. There are at least eight human studies comparing microbiota composition in MS patients and healthy controls. The details of microbes vary between studies, but general themes emerge. There is a decrease in short-chain fatty acid producers including

Faecalibacterium prausnitzii, which decrease the anti-inflammatory effects of butyrate, as well as an increase in pro-inflammatory microbes such as *Enterobacteriaceae*. More recently, *Akkermansia muciniphila* and *Acinetobacter calcoaceticus* were associated with MS patients and shown to cause proinflammatory responses in both human blood and when added to germ-free mice. Further, there was a decrease in *Parabacteroides distasonis*, an anti-inflammatory microbe that is known to induce Treg cells. When feces from MS twin patients were transplanted into germ-free mice, these mice showed more symptoms of multiple sclerosis than those that were transplanted with non-diseased twin human feces. This observation is striking because multiple sclerosis is not an infectious disease—we cannot directly transmit it. So the fact that transplanting microbes can affect immune system function is a big deal. Studies such as these significantly raise the hope that in the near future microbiota can be used not only as a predictor of disease, but potentially as an immune modifier for multiple sclerosis.

Gout

Gout is a type of arthritis characterized by severe pain, redness, and tenderness in the joints. Pain and inflammation result when too much uric acid crystallizes and accumulates in the joints. Consumption of red meat, beer and alcohol, select seafood, and other foods high in purine (which your body converts into uric acid) can contribute to gout. Obesity is a strong risk factor for the disease. Because of rich diet components, gout was historically known as a "rich man's disease" or "disease of kings." Famous gout sufferers include King Henry VIII of England, Benjamin Franklin, Sir Isaac Newton, Ludwig van Beethoven, and Leonardo da Vinci.

But the incidence of gout is increasing today in Western society, probably due to shifts in diet, weight, and longevity. About 4 percent of Americans will have the painful experience of gout in their lifetimes, with symptoms including severe pain, redness, and swelling. During an acute attack, anti-inflammatory medications (NSAIDs) can help to relieve pain and shorten the length of the attack. Behavioural modifications, including diet, exercise, and decreased alcohol consumption can minimize frequency of attacks in chronic sufferers.

A clinical diagnosis of gout is based on measuring uric acid levels in the blood. In a unique application using the microbiome, a group of researchers developed a test based on intestinal microbes. It performed even better than the existing blood test. They hypothesized that intestinal microbes play a role in uric acid metabolism since 30 percent of uric acid is excreted through the intestine. By sampling healthy adults and those with gout, they found that the enzyme that degrades uric acid was lower in those with gout than in the healthy controls (i.e., they would have decreased uric acid degradation, which results in increased disease). The researchers found that a total of seventeen microbial genera were associated with gout and used these microbes to establish a "microbial index of gout." When they tested it on thirty-three healthy and thirty-five diseased people, it had an 88.9 percent accuracy of diagnosing gout, versus a 71.3 percent accuracy for the blood test. This suggests that improved diagnosis and, potentially, treatment of gout could be achieved by using gut microbiota.

Inflammaging: A Key to Aging

Although inflammation is needed to defend against invading pathogens, as is true most of the time, too much of a good thing can be harmful. In the case of chronic low-grade inflammation, we have learned its effects on the long-term health of our tissues and organs, and thereby on how we age. The efficiency of our immune system starts to wane after approximately age fifty, a process called immunosenescence. Although the exact mechanisms of this process are not known, one thought is that long-term overloading with various antigens may in part trigger immunosenescence; i.e., over the years, our bodies get tired of constantly responding to microbial molecules. Furthermore, as we get older, we may not produce enough new immune cells to be able to continue controlling microbes. A decrease in antibody-producing cells

(B cells) and other immune T cells, as well as decreased ability to process antigens, results in a less efficient immune system overall. This is why vaccines do not work as well in older people, and why the elderly succumb to infections such as pneumonia that would otherwise not be as dangerous in young healthy adults.

Although the immune system wanes with age, there is also a marked increase in low-grade inflammation in aging populations. This process is aptly named inflammaging. One thought is that as the immune system's efficiency decreases, it cannot keep the microbiota and other foreign bodies contained as well. These microbes seep into the body to trigger inflammation. As one ages, the gut becomes more permeable. This allows more microbes and their inflammatory-stimulating molecules to enter the body, causing increased inflammation. Such immune stimulators trigger inflammatory responses which, through a complex series of pathways involving an increase in pro-inflammatory cytokines and a decrease in anti-inflammatory cytokines, result in low-grade inflammation. Older people (generally over the age of sixty-five) have two to four times more pro-inflammatory cytokines in their blood than younger populations. Measuring inflammatory cytokines is a very good prediction for determining risk of death within ten years, as well as predicting susceptibility to most age-related diseases.

MYTH: Biological aging and the immune system are unrelated.

FACT: Our immune systems are intimately associated with youthful vigour and vitality. Without an immune system, longevity is not an option. The key to a healthy, longer life is to limit chronic low-grade inflammation, which causes general tissue damage associated with bodily aging.

What is the effect of inflammaging on health and disease? In both mice and fruit flies (the favourite animal models of scientists), limiting such inflammation both maintains healthy gut microbiota and extends the organism's lifespan. Because cytokines circulate throughout the body, inflammaging affects most tissues and organs. It is associated with atherosclerosis (which results in heart attacks and strokes), metabolic diseases of the liver and kidney, obesity and type 2 diabetes, muscle and bone loss, cancer, autoimmune diseases, and neurodegeneration associated with depression and dementia. There are other degrading consequences to the body as this low-grade inflammation causes tissue damage. Inflammaging is at the centre of nearly all degenerative processes associated with aging. The key to a healthy longer life is to limit chronic low-grade inflammation.

Still, we cannot help but question: If inflammaging is so bad for us and our longevity, why did evolution allow it to exist? The simple answer is that because inflammaging mainly occurs later in life, long after we have reproduced, there is no evolutionary pressure to select against it. Evolution has selected for us to reproduce, but how long we live beyond our child-producing years is irrelevant genetically. Further, inflammation is critical to our ability to survive infections and other diseases earlier in life, during our reproductive years. That survival itself is key to longevity. So there is no evolutionary reason to get rid of this protective mechanism, even though it becomes detrimental later in life.

What do microbes have to do with inflammaging? As you probably guessed, everything. As discussed in Chapter 7, scientists are uncovering major differences between the microbiomes of older people, particularly centenarians, and those of younger people. Overall diversity decreases around age sixty-five (less diversity is generally a bad ecological sign), and there is a decrease in the beneficial *Firmicutes* (including *Faecalibacterium prausnitzii*) that lessens inflammation. One of the processes associated with inflammation is the production of toxic reactive oxygen species (called ROS), which increases overall oxygen levels and probably contributes to the decline in oxygen-intolerant *Firmicutes*. There is also a bloom in the pro-inflammatory microbes that produce inflammatory molecules such as LPS. These are facultative anaerobes (meaning they can tolerate oxygen, and hence survive in the modified gut environment) and include Enterobacteriaceae (which includes *E. coli*). Another characteristic associated with aging is increased gut permeability (see Chapter 7). A combination of the gut having fewer beneficial microbes and more proinflammatory ones, along with increased intestinal permeability, collectively results in low-grade inflammation and ultimately tissue damage throughout the body.

INFLAMMAGING: THE BANE OF LONGEVITY

Way back in 1908, Élie Metchnikoff, the famous biologist who is considered the father of immunology, proposed that the gut microbiome was the driver of ill health associated with aging. He also believed that if he used microbes (those now found in probiotics such as lactobacilli), they would restore gut health and subsequently increase longevity and health. He believed it so strongly that he consumed significant quantities of probiotic-rich foods like yogurt and fermented foods. Needless to say, his ideas were severely ridiculed at the time. It has taken over a century to scientifically prove him right.

In a neat set of experiments done at McMaster University, Dr. Dawn Bowdish's team demonstrated that microbes are at the centre of aging and longevity in mice. In older mice (which, like humans, had increased intestinal permeability compared to younger mice), researchers could find gut microbes outside the intestine, as well as their products such as LPS. When they tested old germ-free mice, however, as expected, these products were not found. Germ-free mice are known to live longer than normal mice. They do not develop inflammaging, have less tissue damage, and do not have increased intestinal permeability. Bowdish's team demonstrated that aged regular mice's microbiota, when transplanted into younger germ-free mice, triggered inflammaging and intestinal permeability in the younger mice. They also showed that macrophages from old regular mice were not as effective at engulfing and killing invading microbes when compared to old germ-free mice. The macrophages in old germ-free mice still worked normally and were similar to young mice's macrophages in their ability to kill microbes.

One of the most exciting studies from this team showed that blocking TNF (one of the inflammatory cytokines associated with aging) could reverse microbial dysbiosis. In other words, reducing TNF levels to help limit inflammation also returned the microbes to a healthier balance. Collectively, these experiments suggest that a lifetime of exposure to microbes and their products leads to a gradual increase in intestinal permeability and inflammaging. This also affects microbial dysbiosis and macrophage dysfunction, which in turn lead to more inflammation and subsequent tissue damage.

Knowing that inflammaging is central to the aging process, is there anything we can do to try to counter inflammation at the right time (i.e., when it becomes detrimental in old age, but not when we need it in youth)? At present, we think so but don't yet have any scientific proof. All data suggest that inflammaging could be a reversible process, so there is certainly hope. There are studies underway in older adults following Mediterranean diets versus control diets for a year to see if this reduces inflammaging. Also, if probiotics help to maintain a healthy gut microbiome and decrease intestinal permeability, perhaps probiotics that reduce inflammation and gut permeability may help decrease inflammaging. When we recognize the microbes' role in immune function, as Metchnikoff pointed to over a century ago, we see many novel possibilities to control and reduce inflammaging. Sometimes science takes an awfully long time to get it right.

In This Together: The Cooperative Future of Microbe Immunology

There are two main takeaways from this chapter: (1) the microbes and the immune system are intertwined, and (2) inflammaging is a bane of longevity. We know that the immune system shapes microbial composition in and on our bodies, but we also now realize that the microbes actively shape the immune system. This can affect everything, from controlling infections to triggering autoimmune diseases.

Even though the immune system quiets down in later life (called immunosenescence), microbial products leak into the body to cause low-grade inflammation. Years of inflammaging cause tissue damage, which results in the breakdown of the body's systems. As we understand more about how inflammaging contributes to aging, fascinating new ways of countering this process through the microbes will become available. There is significant hope that we can exploit these new findings to decrease the overall incidence of autoimmune diseases and slow down the natural degradation of body tissues that contributes to illness as we age.

QUICK TIPS

- **Take help when you can:** Keep your vaccinations up to date (including a yearly flu shot each fall) to help your immune system fight off viruses.
- **Age, don't inflammage:** Maintain as robust and diverse a microflora as possible to help diminish gut permeability and low-grade inflammation. Dietary choices to help limit inflammaging include those that increase SCFA (namely fibre) and antioxidants. High-fibre foods include legumes, berries, and cruciferous vegetables. Dark-coloured grapes, blueberries, red berries, nuts, and dark green veggies are examples of antioxidant foods.

12

Flex Your Microbes: The Musculoskeletal System

Weakness. Weight loss. Low energy. Agonizingly slow gait. These are the telltale signs of frailty visible among many elderly people. Contrary to these clichés, however, frailty is a medical condition but *not* an inevitable result of aging. It can be prevented or delayed with deliberate choices that affect lifestyle, behaviour—and your microbiome. Dr. Heather McKay, professor in the Departments of Orthopaedics and Family Practice at the University of British Columbia, is keen to educate people on why these "decline and decay" narratives aren't true. Her research focuses on addressing the needs of our aging population, specifically related to exercise physiology. This stems from her own personal experiences as an elite athlete (she represented Canada as a sprinter on the world stage, including at the Commonwealth Games) and as a mother. When she had children, her experiences showed her it was "really obvious that we set the course of our lives by what we choose to do in young childhood." She set her sights on medical school even though she had young children at home. "But then I realized that I don't care as much about treating people once they're already sick. I want to keep them healthy!" In an abrupt turn away from traditional medical practice focused on acute care and treatment, she became all about prevention.

The need for this kind of focus has never been greater. Dr. McKay acknowledged that she's partially drawn to aging research because of the increase in the population over sixty-five, a group more numerous "than those under the age of fifteen—a first in North American history. If our healthcare system is to have any chance of surviving whatsoever in the future, we need to make sure older adults age well—and age actively." Dr. McKay reflected that as bones grow in children, so do bones decline in older adults: "This is a logical extension of my earlier work. Turns out I've been looking at old bones for a really long time! Understanding how everything is connected across the course of life is important to sustain and maintain skeletons wherever possible. Exercise and diet can play a huge role in the health of our bodies at every age." Dr. McKay is fascinated by longevity and thinking through how we can plan for the final thirty years of our lives, which she calls our "third trimester." "I'm convinced that physical activity and mobility are the keys to a longer life," she said, "and, more importantly, a high quality of life."

She qualified that physical activity does not necessarily mean the blanket prescription of sixty minutes of moderate to vigorous physical activity five days per week. This is just "not in the wheelhouse" of the low-mobility older people she works with. Instead, Dr. McKay takes an incremental approach with light physical activity: "Every little bit counts. If you're lying in bed most of the day, try to sit up at bedside. If you're sitting at bedside, walk to the door. If you can get to the door, walk around the block. Do a little more every day, ideally with friends and family around for encouragement and connection." Even

MYTH: Trying to exercise and get healthy is pointless in later life because decline in old age is inevitable.

FACT: Many of the symptoms that we associate with elderly people—such as weakness and loss of balance—are actually symptoms of inactivity, not old age. You're never "too old" to move your body! On the one extreme, there are body builders and marathon runners in their seventies, eighties, and nineties; on the other, studies show that even nursing home residents in their nineties can boost muscle strength by starting an exercise program.

things like patio gardening and playing with grandchildren count as ways to make people stronger, prevent bone loss, improve balance and coordination, boost mood and memory, and ease the symptoms of numerous chronic conditions.

Fighting Frailty

In 1900, the average American lived to forty-seven years. Nowadays, our average life expectancy is nearly seventy-nine years. Adding thirty years to the human lifespan is an amazing success story, but it also brings new challenges, including frailty—a geriatric syndrome characterized by weakness, weight loss, and low activity. Its two main physiological contributors are loss of bone mass (osteoporosis) and muscle mass (sarcopenia). In the United States, approximately 15 percent of the older non-nursing home population—aged sixty-five and older—are considered frail, meaning that they have three or more of the following symptoms: exhaustion, low physical activity, weakness, slowness, and shrinking; another 45 percent are considered pre-frail (one or two symptoms). In those aged ninety and older, frailty rises to 38 percent of the population. It is estimated that by 2050, 1.2 billion people worldwide will be frail.

Frailty is a strong predictor of disability and multiple adverse health outcomes. A Johns Hopkins study found that frailty doubles the risk of surgical complications, lengthens hospital stays, and increases the odds of moving into nursing homes or assisted living after a surgical procedure by as much as twentyfold. Over half of frail people in the United States have reported a fall in the past year. Over one-third have fallen multiple times, and two-fifths have been hospitalized from falls or other health conditions. Frailty is a more accurate predictor of mortality and hospitalization than actual chronological age. In other words, aging well requires minimizing and hopefully preventing frailty.

What can we do to protect ourselves? Per a major theme in this book, diet and exercise do wonders for a person's general health and longevity—and both involve microbes, as we've discussed. One cause of frailty is the age-related loss of muscle mass, so try to be active and moving most days of the week, choosing activities that can improve strength and thereby reduce weakness. You don't necessarily need to hit the gym to achieve this—as Dr. McKay advised, carrying grocery bags, taking the stairs, and lifting and holding grandchildren all count.

Maintaining a balanced and nourishing diet, including fruit, vegetables, protein, healthy fats, whole grains, and low-fat dairy products, is also important. In one study, people who faithfully adhered to the Mediterranean diet were 74 percent less likely to become frail. Another way microbes are involved in frailty is by reducing nefarious low-grade age-related inflammaging. With age, the number of microbes making anti-inflammatory molecules such as butyrate decreases, while other more inflammatory microbes increase in number. Both of these microbial shifts increase inflammation.

There is also a general decrease in microbial diversity associated with frailty, in addition to the many other diseases discussed throughout this book. It is still not clear whether these microbial differences are due to external changes in diet and housing that accompany aging (see Chapter 13 for more on the role of the environment in later life), or whether they are caused by the aging body's biological changes.

Although causation hasn't been proven, several studies have now examined how the gut microbiome is associated with frailty (microbiomes from other body sites have not yet been examined). In one small study of twenty-three individuals (median age of eighty-six), researchers compared the gut microbiome of ten highly frail individuals to thirteen low-frailty people sharing the same diet and living in the same care home. They found seventeen key gut microbes that showed significant differences between the two groups. Those with high frailty had lower numbers of anaerobic microorganisms lactobacilli, *Bacteroides*, and *Faecalibacterium prausnitzii (F. prausnitzii)*. Remember that these microbes produce butyrate, an anti-inflammatory short-chain fatty acid—so having fewer of these helpful microbes leads to more inflammation. These individuals also had higher numbers of *Enterobacteriaceae*, which are associated with increased inflammation.

In a much larger study of a total of 728 female twins in the UK, researchers sequenced the participants' fecal microbiomes to explore this connection. The average age of participants was sixty-three (ranging from forty-two to eighty-six). Because of the younger ages, many participants still lived in the community and overall frailty scores were lower, yet their results show a similar trend to the smaller study described above. Overall, frailty was associated with a decrease in diversity in the microbiome. Frailty was more strongly correlated with microbial diversity than someone's age, diet, alcohol intake, smoking, or weight. Researchers also found a negative correlation with *Clostridiales*, especially *F. prausnitzii*, a key butyrate producer. This means that frail participants had lower numbers of anti-inflammatory microbes. They found positive correlations between certain other microbes and frailty, although we don't know at this time if they actually increase frailty.

Another major study on the microbiome and frailty was conducted with 178 individuals of Irish descent aged 64 to 102 years old. They used thirteen younger adults with a mean age of thirty-six as controls. One interesting aspect of this study is that researchers included analyses of the microbiome by diet *and* residence: eighty-three lived in the community, while sixty resided in long-term residential care. It took about a year for those entering long-term care to fully shift from a diverse "community dweller" microbiome to a diminished "long-term care" composition (more on this in Chapter 13). Long-term residents had a higher frailty index score, as measured by the five symptoms described earlier—which makes sense since they were in long-term care–and higher levels of circulating inflammatory cytokines such as TNF alpha. Collectively these studies show that we need to keep our microbiomes in mind when it comes to the long-term strength of our bodies.

Brittle Bones

Loss of bone mass is a natural process following the steady increase in bone mass and strength in childhood. There is strong evidence that it is vital for children to do weight-bearing exercises such as jumping and running to build healthy bones that will set them up for life. Jessica remembers her pediatrician mother once instructing her and her brother to go outside before dinner to jump up and down twenty times. At the time, she thought it was just a fun game—but now she sees how her mom was trying to protect her bones!

Dr. McKay's work, and that of others, shows that we gain about 26 percent of our adult bone mass in the two years around puberty. This is as much bone as we lose across fifty years of adult life. Osteoblasts—cells that secrete the matrix for bone formation—build bone, while osteoclasts, cells responsible for bone dissolution and absorption, tear it down. Peak bone mass happens between the age of twenty and the mid-thirties, which is the most bone you will ever have. Then as you age, the bone remodelling process changes: Your body continually builds and breaks down bone structure, replacing about 10 percent of your skeleton every year, but the new bones come in at a slower rate. This can lead to osteopenia: the condition of thin bones, usually a precursor to osteoporosis if not treated properly. Osteoporosis results in significantly weakened bones and increased risk for unexpected fractures.

MYTH: Exercise isn't safe for elderly people, since they are more likely to fall and break a hip.

FACT: Studies show that exercise can actually reduce your chances of falling by building strength, balance, and agility. Gentler activities such as tai chi and yoga may be especially helpful to improve balance and strengthen bones, also reducing risk for—and slowing the progression of—osteoporosis.

Over the age of fifty, the odds of experiencing a major fracture become one in two for women and one in six for men. Post-menopausal women are twice as likely as men of the same age to suffer a fracture, partly due to women's generally lighter frames and smaller, thinner bones, and the hormonal changes that accompany menopause.

There are two types of osteoporotic fractures that are particularly problematic in older people. The first are compression fractures of the spine, which cause the vertebrae to collapse and the spine to curve forward, forming a hump. The abdominal cavity then compresses, which contributes to numerous issues including problems with breathing, digestion, urinary function, mobility, and balance. The second common type of bone fracture occurs at the hip, usually in the narrowest part of the femur bone where it connects to the pelvis (called the femoral neck). These fractures can cause great pain and restricted mobility; typical treatment includes surgical insertion of nails/screws or a partial or total hip replacement to fix the break. In addition to the injury itself and enduring surgery, older people can struggle with the significant and long recovery process. Twenty percent of older patients that suffer a hip fracture die within a year, and 50 percent of those who survive will lose their independence in their living situation and daily activities. Over the course of one's lifetime, the risk of having an osteoporotic fracture is 30 to 40 percent, resulting in major healthcare costs.

MYTH: Only women need to worry about age-related bone loss and osteoporosis.

FACT: Women are four times more likely to get osteoporosis. However, men still make up 20 percent of all cases. More than two million men in the United States have osteoporosis, and another three million have osteopenia.

What causes osteoporosis? The causes are complex and many. It starts in our youth where, depending on genetics (which accounts for as much as 50 to 80 percent of bone mass), childhood activity patterns, and diet, an individual's bone mass, density and strength are first established. Throughout adulthood, changes in, or maintenance of, nutrition and lifestyle factors are important. Diets low in calcium and cigarette-smoking can both weaken bones. Also, weight-bearing physical activity through mechanotransduction—when cells convert mechanical stimulus into electrochemical activity—has direct effects on bone accrual and loss. Because women have longer lifespans than men, they have more time to incur wear and tear on their bones—but that is just part of the reason why women are four times more likely to get the disease (two out of three women over the age of eighty have osteoporosis).

There are hints that the microbes are involved in osteoporosis. First, people with inflammatory bowel disease (IBD) are at a much greater risk for osteoporosis. Second, we learned in Chapter 9 that menopause is characterized by decreased estrogen levels, which contribute to decreased bone mass and density. As estrogen declines, there is an increase in pro-inflammatory hormones produced by T cells, such as TNF-alpha, which cause an increase in osteoclasts (the cells that break down bones). Third, we know that microbes are involved in diet, nutrition, and exercise, all of which can affect osteoporosis. The microbial role in all three factors suggests that bacteria might have to do with development of, or predisposition to, the disease.

Scientists are beginning to formally examine how the microbiome plays a role in osteoporosis with germ-free animals. Data on their bone density and mass are mixed and not at all conclusive. Some studies have shown that a specific breed of germ-free mice (C57B1/6) had higher measures of bone volume than regular mice. When they exposed these germ-free mice to microbes, their bone density decreased to the levels of regular mice. Studies with a different strain of germ-free mice (BALB/c) found the opposite: These mice had weaker bones. Long-term colonization of microbes in certain germ-free mice has resulted in increased bone formation and bone mass. These contrary findings may be explained given different mouse strains, or differences in mouse sex and age.

Last but not least, researchers compared hormone levels and bone mass in conventional and germ-free mice. Using the hormone leuprolide to deplete estrogen in mouse models, they saw that estrogen-depleted germ-free mice did not produce the inflammatory cytokines associated with osteoclast generation and bone loss. After studying the mice for the first four weeks of their lives, researchers colonized the germ-free mice with microbes. This had surprising results: The germ-free mice were *protected* from bone density loss associated with estrogen loss, but the conventional animals had increased bone loss similar to levels seen in post-menopausal women.

The few human studies on the subject found that the microbiome can have a large effect on bone remodelling (the natural processes of osteoblasts and osteoclasts) and bone mass. A small study of eighteen people involved six patients with osteoporosis, six with osteopenia, and six without bone loss (normal controls). Researchers found significant differences in each group's microbiome composition and could differentiate between normal controls and those having bone loss by way of their microbiome diversity levels alone. Surprisingly, it was the normal controls who had less diversity. When Jessica read this result, she had assumed that people without osteoporosis would have more diverse microbiomes. In most cases we've seen in this book, more diversity is generally good—yet this study, like the vaginal microbiome examples, found the opposite. It just goes to show how much we have to learn when it comes to the complex relationships between microbes and disease, and how badly more research is needed to understand claims from studies such as this. Researchers also found several differences in specific microbe genera between normal controls and those with osteoporosis.

Why these unpredictable findings? There are several different ways the microbiome could influence bone density and mass. Inflammation triggers the formation of inflammatory cytokines, some of which trigger the production of those bone-breaking osteoclasts. We know that menopausal women and IBD patients, as well as older people, have higher levels of inflammatory cytokines due to microbial activity, which may further activate osteoclasts.

Calcium absorption is another factor that plays a role in bone density, and several studies find some evidence that the microbiome is involved. When gut microbiota break down dietary fibre consumed in foods, and produce acids such as SCFAs, the gut pH decreases; and lower pH boosts calcium absorption into the body. These SCFAs, especially butyrate, can also modulate signal pathways in the gut, which can increase calcium absorption and affect bone density.

Some of the best evidence that the microbiome affects bone density comes from research with probiotics and prebiotics. Several mouse studies show that modulation of the gut microbiome by probiotics (mainly lactobacilli) increases bone mass and can prevent estrogen-mediated bone loss. There is also data coming out about humans. In a study of the probiotic *Lactobacillus reuteri*, researchers found that it increased circulating levels of vitamin D, which promotes calcium absorption and bone health. There was also a large study of 417 older patients with broken arms, some of whom researchers gave the probiotic *Lactobacillus casei*. They found that fractures healed faster among those who received the probiotic than those who did not. It has also been shown that women with osteopenia who received a probiotic mix of six microbes for six months showed decreases in the inflammatory cytokine TNF-alpha. They had increased markers of bone formation, but researchers didn't see any effect on bone density during this short period of time.

These findings usher in a new area of research generating significant excitement. While conclusive data is yet to come on the many linkages between the microbiome and bone density, measures like taking probiotics, enhancing microbial diversity through dietary sources and, of course, maintaining a calcium-rich diet are all recommended to reduce risk for osteoporosis—and to improve health in general.

Microbes with Your Muscle Milk

Along with osteopenia and osteoporosis, muscle loss, called sarcopenia, plays a major role in frailty. The diagnosis refers to generalized muscle loss, which includes both muscle mass and strength. It is associated with physical disability, heightened risk of falling, and, ultimately, increased risk of death. Sarcopenia generally accelerates around the age of seventy-five, although it can vary in intensity from ages sixty-five to eighty to increase one's overall frailty. Several factors account for sarcopenia, including a decrease in nerve cells connected to muscles, changes in hormones such as growth hormones and testosterone, and decreases in protein intake and synthesis.

Why do we lose muscle in the first place? Muscle mass peaks later than bone mass, at around the age of thirty. With age, the balance between muscle growth and loss shifts, and more muscle is lost than synthesized, resulting in a decline in both muscle mass and function. Physically inactive people can lose 3 to 5 percent of their muscle mass every decade. That's up to thirty percent of muscle mass loss between the ages of twenty and eighty! The primary treatment for sarcopenia is resistance or weight training to increase muscle strength and endurance. These workouts differ from cardiovascular exercise, although they often overlap. Several studies show significant benefits within two weeks of starting training. Continued exercise helps prevent further muscle loss and can even build some back, but it can't totally prevent it.

MYTH: You cannot gain muscle from a solely plant-based diet.

FACT: You can maintain strong bones and muscles as a vegan or vegetarian. What's key is getting adequate protein through supplements and/or foods consumed (e.g., nuts, beans, legumes, leafy greens, non-dairy milk, cruciferous vegetables, seeds).

The links between sarcopenia and the microbiome are only beginning to emerge. In a mouse model of leukemia—which features muscle atrophy—there was a correlation with changes in the gut microbiome including reduced *Lactobacillus*. By adding back specific species of lactobacilli, muscle atrophy in this mouse model was reduced. Microbiome studies have also been conducted on aging rats (who undergo sarcopenia, like humans). Comparing the gut microbiome to muscle physiology, researchers found specific age-related changes in the microbiome that correlated with the physiological decline of musculoskeletal function.

We view this result cautiously because, as we saw in Chapter 7, there is a general change in the gut microbiome as one ages. Whether this change is causing musculoskeletal changes or is just correlated with it remains to be determined.

There are no studies yet in humans that directly examine the effect of the microbiome on sarcopenia. However, one study indicates there may be an association through diet. As part of the TwinsUK study, 2,750 women between the ages of eighteen and seventy-nine were asked to follow a Mediterranean diet. They were then scored for their adherence to that diet. Results reflected a direct positive correlation between higher muscle mass and strength with higher adherence to the diet. Interestingly, there was an inverse correlation between meat consumption and muscle strength. In other words, those who ate more meat had less muscle strength (which further disproves the common myth that you can't build muscle on a plant-based diet). The dietary effect was greater in women over fifty, which is also when there is increased sarcopenia. This is the first study on sarcopenia to look at diet across ages, rather than only in elderly populations, and it suggests that how we treat our microbiome throughout life can affect our health as we age.

MYTH: If you're not working up a sweat, you're not working hard enough.

FACT: Sweat is not necessarily an indicator of exertion—it's the body's way of cooling itself. Heart rate is a better indicator of exertion for cardiovascular exercise. In weight training, you may not necessarily work up a sweat depending on the type of exercise, your body's propensity to sweat, and temperature of the room.

Feel the Burn

There are tantalizing hints from recent human studies that exercise influentially boosts the microbiome, as we touched on in Chapter 8, with an overarching theme: Exercise increases microbial diversity because it selects for helpful SCFA producers, which can decrease inflammatory responses and facilitate overall health.

Several studies in rats and mice provided early evidence of the effect of exercise on the gut microbiome. A study of rats placed in cages with running wheels for "voluntary exercise" found that these rats had higher levels of the SCFA butyrate in the colon compared to sedentary animals with no access to a running wheel. These exercisers also had an increase in beneficial butyrate-producing firmicutes microbes in their intestine. In another mouse study, researchers compared three groups of mice: germ-free animals, sedentary normal animals (no running wheel available), and an exercise group (running wheel in their cages). As other studies support, the data here showed that exercise increased microbial diversity and butyrate-producing microbes, which translated into decreased inflammation. Researchers then transplanted the microbiota from the exercised and sedentary mice into the germ-free animals; the transplanted microbiota from exercised mice decreased inflammation in a mouse model of IBD. Collectively, these studies show that exercise alone may modulate microbial composition and cause changes in the microbiome to decrease gut inflammation.

Only one mouse study that we know of thus far has examined the reverse scenario: How the microbiome affects exercise performance. This study tested endurance during swimming of normal mice, germ-free mice, and germ-free mice colonized with a single bacterium (*Bacteroides fragilis*). It found that mice colonized with normal microbes or even the single bacterium were able to swim for longer times than germ-free animals. This suggests that the microbiome might affect exercise performance. This is probably not surprising, since we know that the microbiome can affect energy metabolism, immune responses, and stress responses—all of which play a role in exercise performance.

Although there are only a few such studies so far, a similar trend appears to hold true in humans. One fascinating study looked at the changes in eighteen lean and fourteen obese sedentary individuals when they began exercising. They started with thirty minutes of moderate exercise (walking) three times a week, and over six weeks progressed to sixty minutes of vigorous exercise (vigorous jogging or cycling) three times a week. They did not change their diets. Participants then stopped the exercise program and researchers tracked them for another six weeks in the "washout period." As in the animal studies, researchers saw an increase in microbial diversity and microbes that produce SCFAs for all exercisers, although the changes were more pronounced in the lean individuals. After stopping the exercise program, the microbiome reverted back to its original composition.

In a 2017 study of women, moderate physical exercise was shown to modify the composition of the human microbiome and increase the abundance of health-promoting bacteria such as *Bifidobacterium*, *Akkermansia*, and *F. prausnitzii*. All of these microbes are associated with SCFA production and anti-inflammatory activity. Similarly, an analysis of 1,493 participants in the American Gut project showed that increasing moderate exercise boosted microbial diversity in the *Firmicutes*, including the butyrate producer *F. prausnitzii*. Physical exercise is commonly recommended to prevent and treat chronic inflammation in many diseases such as type 2 diabetes, coronary disease, and obesity, all of which are characterized by dysbiotic microbiota. These exercise studies collectively suggest that moderate exercise increases the levels of anti-inflammatory SCFAs by beneficial modulation of the microbiome.

BUG DOPING

Given all that we are learning about the role of the microbiome in exercise, it's possible that we might be able to design the perfect (and presumably legal) microbiome to enhance the performance of elite athletes. There is only one study thus far that has focused on elite athletes' microbiome: that of Irish rugby players (described in Chapter 8). In our eyes, this is a prime area for stealth scientific innovation. Here we describe the ingredients of our *hypothetical* microbial cocktail.

- First, we know that elite athletes' training habits differ from most "weekend warriors" or "hobby joggers." They train frequently (often two to three times per day) and repeatedly find creative ways to push themselves to the limits to meet specific goals. Extreme exercise triggers significant inflammation in the body, which includes inflammatory cytokines. Athletes generally recover quickly and their levels of inflammation drop back rapidly to very low levels. So to design our perfect bug doping concoction, we first need to include microbes that produce anti-inflammatories (namely the butyrate producers we discussed above) to even further enhance this response.
- Second, we need our microbial elixir to include a mix of microbes that break down lactic acid, which is a by-product of intense exercise. Think of that nauseous feeling you got in elementary school after sprint repeats in gym class, or if you push yourself through gruelling high-impact interval training. Lactic acid can limit performance when oxygen cannot reach the muscles as efficiently, and energy breakdown (metabolism) switches to pathways that don't use oxygen (anaerobic metabolism). One can recover upon decreasing exercise intensity or resting, which returns the metabolism to its aerobic state (using oxygen), and the blood circulation washes away the lactic acid. In one study looking at twenty runners training for, and running, the Boston marathon, researchers found that post-race runners had a major enrichment in a microbe that breaks down lactic acid. Jessica ran the 2018 Boston marathon and can personally attest to gruelling long runs and tough workouts that likely selected for this microbe.
- Third, we'd need microbes that could efficiently break down food to rapidly generate energy during a race and increase protein uptake to help repair muscles. This is especially important for the ultramarathoners tackling fifty- to one-hundred-plus-mile races. Although scientists have found microbes in these athletes that

efficiently and selectively break down carbohydrates and fibre, applying that knowledge broadly will be a bit tricky until optimized personal diets gain more headway.

- While we are dreaming, we may as well throw in some microbes that affect the brain. Perhaps we can find microbes that dampen pain, which would help us push through soreness and fatigue. As we saw in Chapter 3, we can probably dig up some microbes that overcome stress too, and alleviate symptoms for athletes who suffer from crippling pre-race nerves at the starting gate. And as we have seen in this chapter, let's throw in some microbes to strengthen our bones and muscles to reduce injury and fractures. Gymnasts might especially benefit from having stronger muscles to ensure a perfect landing every time.
- Finally, let's tackle jet lag, a major problem for international athletes. We can likely bring in microbes that affect circadian rhythm to regulate their bodies' clocks and keep them performing at their best in any time zone. We also have to remember that elite athletes are only human, so they still get colds and other communicable diseases—and could be more susceptible while travelling on planes or running themselves ragged with training. Imagine the devastation of getting sick while competing at the Olympics! So, let's throw in some beneficial oral microbes that outcompete respiratory viruses, and for good measure, wash it down with probiotic yogurt. In forty-six female endurance swimmer athletes, probiotic yogurt was linked to a reduction in the number of respiratory infections and some of the symptoms.

A final touch of this hypothetical world would call for some creative coaching strategies: Take the very best person in your sport—a multi-gold medal Olympian and world champion will do quite nicely—and pay them an exorbitant amount of money for their feces, which they'd just flush away anyway (so what's the loss?), then line up medical fecal transfers for your athletes to receive a dose of microbes from this individual. At present, there is no regulation regarding athletes' microbiomes—so the biggest hurdle might be getting the athletes to cooperate in this unorthodox training technique!

QUICK TIPS

- **No one is "too old" to move:** A mix of physical activities that are weight-bearing (these can be short and sharp like jumping), make your heart beat faster (this can be short and fast high-impact training, or long and slow like walking or running), and increase your strength (weight training) are important at any age. Finding a functional and enjoyable exercise routine is an important way to get regular bursts of activity instead of sporadic sweat sessions at the gym. Go for a walk outside, play with grandchildren, go bowling with friends, dance in the kitchen while cooking—whatever you enjoy. As Dr. McKay recommends, small bouts of exercise that help you sneak in physical activity have cumulative benefits in strengthening your bones and muscles—and in the process, boosting your microbes as well.
- **Eat your veggies:** Plant-based foods help boost dietary fibre intake, which can increase calcium absorption through the gut microbiota. Choose fibre-rich foods such as beans and legumes, whole grains, brown rice,

nuts, baked potatoes with skin, berries, bran cereal, oatmeal, and vegetables—all staple components of the Mediterranean and MIND diets (see Chapter 3). You can double down by serving these foods with calcium-rich dairy products, such as bran cereal with milk and baked potatoes with sour cream, to further strengthen your bones and reduce risk for osteoporosis.

13

Too Clean, or Not Too Clean: Environmental Microbes

One number predicts your health and life expectancy more than any other: your zip code. Environmental health researchers and epidemiologists have long recognized the importance of geography to human well-being. This includes physical aspects of the places we live, learn, work, and play (the built environment); the people we interact with (the social environment); and local characteristics of the natural environment (such as air and water quality and proximity of green space). Having access to healthy foods, quality healthcare services, and well-maintained parks; ingesting clean water, food, and air; spending time with people who nurture and support us; and living in safe, clean homes all have the potential to influence our overall lifestyles and health.

When we zoom into these various contexts, we once again discover the partners that are silently conducting the symphony of our world: microbes. Microorganisms are present in all these places—the indoor and outdoor spaces we live in, the food and drink we consume, the people we're with every day or only pass by on occasion—and are therefore a key puzzle piece in explaining why some people are healthier and live longer than others. Older adults are particularly susceptible to the conditions of their homes and neighbourhoods, given that they may be less mobile and spend more time in restricted locales—and as we saw in Chapter 11, may have a compromised immune system for fighting off potential dangers.

Jessica focuses her research on this complex relationship between environments and aging. As an environmental gerontologist, she studies where people live as they age and how that affects their health and well-being. Her work examines how environments can "get under our skin" to shape experiences of aging.

When interviewing older adults across the Minneapolis metropolitan area as part of her PhD research, Jessica could not help but notice drastic differences in people's microbial exposures. Some homes were spotless and sterile with freshly scrubbed counters and floors smelling strongly of bleach. More often, homes featured piles of belongings—from clothing to food containers to paperwork–multiple human and pet inhabitants, and evidence of many years of residence. On occasion Jessica even observed mold from water damage, rotting food on counters, and stuffy unventilated air. This likely put inhabitants at risk through regular contact with pathogens. With this clear-eyed and more holistic view of the aging process, she began to consider invisible and dynamic interactions between the environment, microbiota, and older people.

Nature Versus Nurture

The role of the environment in microbial composition is a subject of much scientific debate in the field of microbiology. To investigate this topic, Dr. Eran Segal and colleagues from the Weizmann Institute of Science collected blood and stool samples from 1,046 healthy Israeli adults (a 2018 study published in *Nature*). These individuals were aged eighteen to seventy; had not used antibiotics within three months prior to participation; did not have a chronic health condition, and were not pregnant; and were from five distinct ancestral origins: Ashkenazi,

MYTH: Longevity is determined by genetic factors.

FACT: The everyday social and physical conditions of life—where you live, who you interact with, what you eat, how often you exercise—can influence health and longevity to a much greater extent (about 75 percent in studies) than genes (about 25 percent).

North African, Yemenite, Sephardi, and Middle Eastern. The relatively recent immigration of genetically diverse populations to Israel, where they share a common environment, created ideal conditions to compare the degree to which environments and genetics shape the microbiome. Segal's group found that ancestry was not significantly associated with microbiome composition; an individual's genetics actually determined a very small fraction (less than 2 percent) of the differences seen across the microbiomes of people.

To test this further, Segal and colleagues then analyzed an existing data set from a 2016 study of microbiome composition in 1,126 pairs of twins from the United Kingdom. They determined that between 1.9 and 8.1 percent of the human microbiome is heritable. So even though twins share the same genetics, they still develop into markedly distinct individuals. Lifestyle factors such as diet, drugs, and home environment account for the remaining 91.9 to 98.1 percent.

In order to examine the degree of environmental influence on the microbiome, Segal's group also looked at the microbial compositions of twenty-four pairs of related individuals who had never lived together. They found no evidence of similar microbiomes among these pairs. By contrast, when investigating fifty-five related pairs with a history of household sharing, they found significant similarities in their microbiomes. They also performed analyses on thirty-two pairs of genetically unrelated pairs who had shared a household, and again found significant microbial similarities. The results of this seminal research all suggest that past or present household sharing contributes to gut microbiome composition, given that relatives with no past household sharing do not have similar microbiomes. In other words, the places you inhabit, and not your genes, define which microbes inhabit *you*.

Segal's and others' groundbreaking work demonstrates the value of examining non-genetic determinants of health and disease. When considering these factors, new ways to personalize medical treatments and approaches to vitality and healthy aging emerge. Relative to their genes, people's microbiomes are easy to change. We shall see in this chapter that environmental microbiome interventions and treatments, big and small, long-term and short-term, are poised to help us live longer and healthier lives.

Underexposed and Overkilled

Every time you open the front door, a gust of air carries microbes into your home—as do dogs, visitors, Amazon packages, and your own body, which picked up microbes between your car, the sidewalk, your office, etc. These microbes are critical, as Americans now spend about 90 percent of their time indoors. This goes against human nature. We did not evolve in sterile chambers or closed rooms, but rather in close contact with nature. We slept on cave floors, had limited hygiene rituals, and spent most of our time outside with other people and animals. Even as society progressed, livestock ambled the streets, roofs and walls leaked, sewers overflowed, and windows opened, letting all those microbes in. Until relatively recently, most families lived on farms and were exposed to abundant microbes through daily chores, frequent romps outside, and breezy open windows.

> **MYTH:** All fungi and bacteria in the home are harmful.
>
> **FACT:** This common misconception inspires unnecessary fears and often excessive cleaning, sterilization, and use of antimicrobial products. The vast majority of microbes found in the built environment are innocuous, and many are even beneficial to human inhabitants.

Modernization sealed us away. This means that we are generally exposed to significantly less diverse microbes on a daily basis. On top of that, we now inhabit a world with thousands of antimicrobial products, ranging from paint and carpeting to cutting boards (more on this later). So in addition to being exposed to less microbes, we're also regularly attempting to kill the ones we actually do come into contact with. We may be more productive, connected, and technologically advanced compared to our cavemen and cavewomen ancestors, but we've also lost contact with many of the essential microbes that evolved with us. This shift to the sterilized indoors has had some unintended consequences.

It appears that fundamental changes in lifestyle from farm to urban living have led to decreased contact with certain microbes that are essential to immune system development. One result is that respiratory diseases are on the rise: three hundred million people worldwide suffer from asthma, and more than 40 percent of the developed world's population has allergies. Growing scientific evidence links environmental factors known to provide microbial exposure to reduced risk of developing asthma. In one example, children who grew up on Bavarian or Amish farms, and in close proximity to livestock, had significantly lower rates of asthma than the general, non-farm population. Several factors have been associated with lower prevalence of some allergies, including contact with animals as a child, exposure to stables under the age of one, breast feeding, vaginal birth, and lack of antibiotics early in life.

Being removed from the microbes we knew from farming days doesn't mean that we are alone in our modern-day indoor environments, however. Rather, our homes, work places, and public buildings provide new habitats and comfy residence to numerous microbial communities. The problem is that these resident microbes are generally not as diverse or beneficial in composition as outdoor environments. It is therefore important to get outside at any age. Older adults are too often confined to sterile, climate-controlled indoor environments.

We can also find ways to bring the outside in, such as ventilating with fresh air and surrounding ourselves with live greenery. Indoor houseplants and flowers can positively influence health by introducing beneficial bacteria into the indoor microbiome, in addition to stress reduction, boosted creativity, and extra oxygen production. Just like humans, plants harbour distinct microbiomes. Houseplants can also remarkably improve indoor air quality. Leaves filter air to reduce carbon dioxide levels and release oxygen. Indoor plants can stabilize the ecosystem and even counteract pathogens. For example, English ivy, the evergreen vine we so often see climbing the sides of stately buildings, can help eliminate airborne mold spores. So enhance the biodiversity and air quality of your home through houseplants and flowers. Open the windows when the weather permits to ventilate with outdoor air and let in additional plant-associated microbes.

MOISTURE AND MOLD: YOUR UNWANTED ROOMMATES

Fungal growth in damp or water-damaged buildings is an increasing problem. Researchers estimate that dampness is an issue in 15 to 40 percent of North American and Northern European homes and, according to the WHO, it is present in 10 to 50 percent of buildings in Australia, Europe, India, Japan, and North America. Most indoor water damage is caused by natural disasters (floods) or humans (disrepair). Water can also penetrate buildings through melting snow, heavy rains, or sewer system overflow; water vapour can be produced through daily activities such as cooking, doing laundry, and showering. Mould—a common type of fungus—will easily grow in any place with a lot of moisture.

Most moulds are not harmful to healthy humans. They can, however, cause nasal congestion, throat irritation, coughing or wheezing, eye irritation, and sometimes skin irritation in sensitive people. Long-term exposure can particularly affect older adults; infants and children; individuals with respiratory conditions, asthma, and/or allergies; and immune-compromised people. These groups are at increased risk for respiratory problems and infections, and the exacerbation of asthma and allergies.

The key to minimize mould inside the home is moisture control. The CDC recommends using air conditioners and dehumidifiers to reduce high humidity levels; fixing leaky roofs, windows, and pipes; conducting a thorough cleaning and drying after flooding; and ventilating bathing, laundry, and cooking areas. Mould growth can be scrubbed away on hard surfaces with commercial products, soap and water, or bleach solutions. Most importantly, fix the root cause of any water/moisture problems so that the mould can't grow back.

We Are Family

With scientific advancements in molecular sequencing, we are increasingly able to analyze the individual bacterial, archaeal (a different type of microbe), and fungal species inhabiting the spaces we live, work, and play. We've learned that indoor microbiomes originate primarily from human skin, pets, and the outside air (see Figure 1). Human depositions of microorganisms have been observed in the microbiome environments of classrooms, households, and athletic spaces. Human occupants can deposit skin microorganisms at a rate of over one hundred airborne microbial cells per hour. These microbes can then decay, only to be replaced at a rapid rate on the surfaces frequently in contact with humans. Microbes most commonly shed by humans come from skin bacteria, including high counts of *Actinobacteria* and *Firmicutes*, in addition to skin-associated yeasts such as *Malassezia*. While we normally live in peaceful co-existence with *Malassezia* yeasts (they are integral components of the skin microbiota), an overabundance can cause skin conditions such as dandruff and atopic eczema, as we saw in Chapter 2.

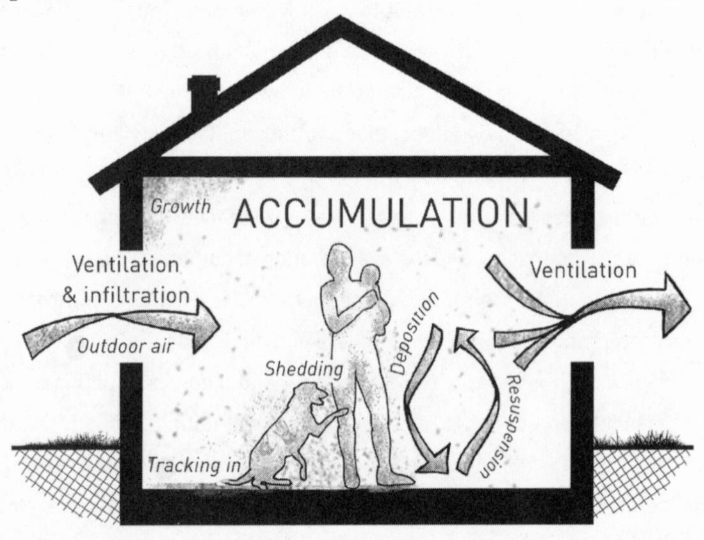

Figure 1: Sources and physical processes that govern assembly of indoor microbial communities. Other potential sources of microbes may include emission from plants, food, and plumbing. Illustration by Sarah E. Kwan and Jordan Peccia (2016). Buildings, Beneficial Microbes, and Health. *Trends in Microbiology, (24)*8. Reprinted with permission.

Scientists in the Home Microbiome Project sequenced bacteria from seven different families, including their pets and shared homes, over four to six weeks. Eighteen participants were trained to collect 1,625 microbial samples from their bodies and from their homes consisting of ten houses, three dogs, and one cat. For three families, samples were taken immediately before and after moving to a new home. Participants swabbed the surfaces of skin, hands, feet, noses, countertops, doorknobs, and many other surfaces that the residents touched in their homes. Hands were the most microbially similar among family members, while noses remained the most distinct. By contrast, microbial communities differed significantly among homes, and the home microbiome was largely sourced from the humans living inside. Turns out, our bodies release bacteria in almost every encounter we have with our environments—when we scratch our heads, yawn, open the fridge door, or flop onto the couch when we come home at night. Our germ-sharing also happens rapidly. When three of the families in the study moved to a new home, it took less than twenty-four hours for the new places to look exactly like their old ones—at least when it came to their microbial roommates. This held true even when the new place was a hotel room that had just been inhabited by other people! Rapid and complete colonization of microbiota in each home was so exact and unique that a swab of any room revealed the family's distinct microbial signature. This was so specific that the researchers were able to successfully predict which family was responsible for a particular sample based solely upon the set of floor microbes they were given.

Social interactions outside our main home also play a major role in facilitating microbe exchanges. Simple acts like shaking hands, passing a friend's phone and scrolling through her photos, and eating off the same dishes all involve swapping microbes. In one study, a volunteer (intentionally) touched a virus-contaminated door handle, then shook hands with a second participant, who then shook hands with a third participant, and so on. The results showed that the virus could be transmitted this way as many as six times. After this person-to-person exchange, the microbes can also be deposited on various surfaces by one person, only to be collected afterwards by another. In one study of ten desk surfaces spread across three schools, students deposited microbes from their skin, oral, and gut microbiomes. After desk cleaning, the microbial communities fully reestablished themselves on the surfaces within two to five days.

Both young children and older adults concerned about being more susceptible to infections should wash their hands regularly with warm, soapy water. Be cognizant of the many people and surfaces you swap microbes with on a daily basis.

CHECK YOUR BACTERIAL BAGGAGE

Lifestyle changes and travel shift our microbiomes in response to the new people, places, and foods we encounter. One of the worst fears of exotic travel is experiencing traveller's diarrhea (TD) or other enteric (intestinal) infections. In one study, a participant relocated from a major American metropolitan area to the capital of a developing nation in Southeast Asia. This individual was exposed to a novel diet and environment while travelling, and experienced two bouts of diarrhea. He also experienced nearly a twofold increase in the bacteroidetes to *Firmicutes* ratio, which reversed upon his return to the United States.

The human gut microbiome plays a major role in travellers' health. The density and diversity of one's gut microbiome can prevent and limit pathogenic colonization and growth. This occurs mainly through competitive exclusion. A traveller's unique gut microbiome may predispose him or her to enteric infection to a greater or lesser extent. For example, Swedish tourists travelling to high-risk destinations for enteric infection were more susceptible to contracting *Campylobacter* (a cause of diarrhea) if their pre-travel microbiome had a lower diversity. The structure and composition of the gut microbiome can protect against common traveller gut ailments. In another study comparing the gut microbiomes of individuals with TD, researchers found a dysbiotic profile of a high *Firmicutes* to bacteroidetes ratio in travellers who developed diarrhea.

There may be ways to manipulate the microbiome to prevent TD. One option is to maintain as normal a diet as possible when abroad. Even short-term changes in eating patterns can have long-term effects on your microbiome—even after returning from travel. However, we recognize that this strategy may notably diminish the quality of one's travels in search of new foods and unique experiences.

Similar to how we get shots to boost our immunity to infectious agents before exotic travel, we might also consider arming our guts with helpful microbes. Studies have shown that probiotics including *Saccharomyces boulardii* and *Lactobacillus* GG can play a role in the prevention of TD. Prebiotics including a galactooligosaccharide mixture in travellers was linked to reduced bouts of diarrhea, but it was a small study and not a clinically controlled blind trial. Researchers are still investigating whether probiotics and prebiotics can treat acute TD.

Elderly Institutional Care

Relocating to a long-term care facility or nursing home also has major effects upon the microbiota. If someone moves to a nursing home, it takes about a year for his or her gut microbes to fully shift to the resident local microbiota. Nursing homes often feature an unhealthy microbial composition that increases inflammation due to lack of microbial diversity. Bacterial infections are also common, given that these facilities provide an ideal environment for the acquisition and spread of infection: Residents share sources of air, food, and healthcare in often-crowded institutional settings, and are already more susceptible to disease. Furthermore, the constant flux of visitors, staff, and residents brings in pathogens from both the hospital and community. Outbreaks of respiratory and gastrointestinal infections are frequent in these settings, and as we know from other chapters, they are caused by microbes and have strong microbial links regarding disease susceptibility.

The bacterium *Streptococcus pneumoniae* (often called pneumococcus) causes many common infections including pneumonia and ear infections. It is a leading cause of illness and death in the United States each year. We generally use antibiotics to combat bacterial infections like those caused by the pneumococcus. The issue is that because antibiotic usage is so common—and sometimes inappropriately prescribed—in both young and (especially) old populations, care environments often become ideal settings for drug-resistant strains to emerge. The CDC reported that between 10 and 40 percent of pneumococcus infections are drug resistant, and these rates are on the rise. Drug-resistant outbreaks can be prevented or minimized with careful use of antibiotics, and pneumococcus pneumonias can be prevented through vaccination. The issue is that vaccines are often underused in nursing homes. Physicians may mistakenly believe that vaccinations are ineffective or harmful in older populations. Only 45 percent of people aged sixty-five and older have been vaccinated in the United States, and in most nursing homes with outbreaks, fewer than five percent of residents have been vaccinated.

A 2017 meta-analysis published in the *American Journal of Infection Control* stated that, on average, 27 percent of nursing home residents are colonized with drug-resistant bacteria. Risk factors included advanced age, comorbid (having two or more) chronic diseases, history of recurrent hospitalization, increased interaction with healthcare workers, frequent antimicrobial exposure, decreased functional status, advanced dementia, immobility, fecal incontinence, and residency in a long-term care facility. The authors underscore the importance of strong infection prevention programs in nursing homes and long-term care facilities. This is a challenge given frequent understaffing, minimal resources, insufficient training, and inadequate surveillance. The hospital environment also falls under consideration here, given frequent transfers of residents between nursing homes and acute care, which contributes to the influx of pathogens between them.

The Hospital Microbiome study began collecting microbial samples from surfaces, air, staff, and patients of the University of Chicago's new hospital pavilion in 2013. Their goal is to better understand the factors that influence bacterial population development in healthcare settings. Just as other researchers studied people moving into a new home, this study examines the rate and structure of succession in a hospital microbiome as the hospital starts accepting staff and patients. At the time of writing the results are not yet published, but we can expect to see environmental variables such as building materials, temperature, humidity, HVAC, and cleaning procedures/schedules playing a role. These factors intersect with the patients and staff to affect the complex microbiomes in the hospital. Observing patterns and trends in this sort of data could be critical to the prevention of people becoming sicker in a hospital—the place where they should be recovering—and could also affect the designs of hospitals and medical facilities in the future.

MYTH: Opening the window in a nursing home or hospital room will let in harmful bacteria.

FACT: Healthcare centres might help thwart the spread of infection by opening their windows. Fresh air lets more microbes in, which can help control previously unchallenged pathogens. In rooms sealed from the outdoors and unventilated by fresh air, harmful bacteria have less competition from other microbes in the environment. A diverse environmental microbiome can act as a protective layer against hospital and nursing home borne infections.

Fido and Her Microbes

One way to bolster the immune system is through exposure to a wider array of the microbial universe, thereby increasing microbial diversity. Remember, more diversity in the biological world is nearly always preferred! One way to do that is to get a pet—especially a dog—to bring more abundant and diverse microbes into the home. In one study, households with pets had more plant and soil bacteria in their homes, which is usually a good thing. Feeding mice dust from dog-dwelling homes made them more resilient to allergens. And as we've seen, exposure to animals during childhood—including domesticated pets—has been associated with lower prevalence of asthma and allergies. One does not necessarily need to move to a farm, given that dogs can have similar protective effects on their owners as horses (though farms overall likely provide more diverse microbial exposure than city living). Additional benefits of a having a dog in later life include companionship, stress reduction, and an excuse for regular outdoor exercise.

In some cases, cat bites (generally to the hand) cause serious bacterial infections, as cats' sharp teeth make relatively deep puncture wounds and can introduce infectious microbes into the body. We imagine this may be more severe in older people with thinner and more fragile skin. Bacteria usually invade from the oral microbiota of the biting animal but can even originate from the victim's own skin or physical environment at the time of injury. If bitten, be alert for signs of infection. Seek medical care if you develop swelling, redness, pain, or have difficulty moving your hand.

MYTH: A dog's mouth is cleaner than a person's mouth.

FACT: A dog's mouth likely contains fewer microbes that are directly harmful to humans. However, that does not mean that the dog's mouth has fewer microbes overall or is "cleaner." Think about how dogs behave—their mouths frequently come into contact with their own and other animals' backsides and feces (unlike humans, they are coprophagic).

On rare occasion, pets can bring detrimental microbes into the home. *Toxoplasma gondii* (*T. gondii*), for example, is a neurological parasite that can invade all warm-blooded animals including humans. In immunosuppressed individuals and during pregnancy, its effects can be devastating. Symptoms include aggression, impulsivity, severe mental illness, and even suicide. One can get an infection from *T. gondii* from contact with cat fecal matter. This is why pregnant women are advised to avoid exposure to household cat litter. Also, dogs can carry it inside on their fur after rolling in cat feces and contaminated soil. If concerned about a furry friend's hygiene, washing your hands after touching them is a good idea (as we well know by now).

YOUR CELL PHONE IS TEN TIMES DIRTIER THAN A TOILET SEAT

We are constantly connected with our phones, from reading the news during our morning commute to scrolling through music waiting at the doctor's office; making lists and searching for recipes at the grocery store; and even surfing social media in the bedroom. We often don't give it a second thought. Mobile phones are the most frequently owned and used electronic devises worldwide and bear the brunt of much negative media for their impact on our mental health and relationships. And now, research shows that cell phones are far dirtier than we might ever expect. They are actually large reservoirs for bacteria—the more microbes they collect, the more microbes wind up on your hands, face, and everywhere in between.

The text-cramp in your thumb isn't the only way your phone and hand have an unhealthy relationship. Your hands are the biggest culprit when it comes to depositing microbes on the phone's surface. One study showed that Americans check their phones on average forty-seven times per day, which provides plenty of opportunities for microbes to move from your fingers and palm to your phone. Research varies on exactly how many and what kinds of bacteria are inhabiting the average cell phone. A 2017 study found a median of 17,032 bacterial genes per phone among secondary school students' mobile phones. Most organisms were harmless, including those passed through contact from human skin; potentially pathogenic microbes, however, including *Staphylococcus aureus*, were among the dominant microbes found on these phones. No differences were found based on the phone owner's gender or between phone types (touch screen versus keypad). No antibiotic-resistant microbes were detected.

Researchers at the University of Arizona found that cell phones carry ten times more bacteria than most toilet seats. Ironically, unlike the bathroom, people rarely clean or disinfect their phones, so the bacteria just keep building up. As we use our phones frequently, they remain warm, which creates ideal breeding grounds for bacteria. Phones might represent an unknown biohazard to older adults and immune-compromised individuals susceptible to lurking pathogens. Fortunately, there are easy ways to minimize contamination. First, do not bring in or use a phone while in the bathroom. When toilets flush, they can spray up bacteria into the air, which could contribute to how phones end up with fecal bacteria such as *E. coli*. It may be a good idea to put the lid down in general when flushing to prevent fecal matter from spraying onto other bathroom surfaces—like your toothbrush.

There are a few effective ways to clean your phone. A soft microfibre cloth all on its own will remove many of the surface microbes. For a deeper clean, researchers suggest using a combination of 60 percent water and 40 percent rubbing alcohol. Dip a cloth in this solution before gently wiping it across the phone. Cleaning your phone several times a month is ideal. Or just get used to saying hello to your microbes each time you pick it up!

A Kitchen's Dirty Secrets

Exposure to foodborne microorganisms (both harmful and harmless) during food handling in the home is fairly frequent. *Salmonella* is a bacteria that frequently causes foodborne illness, often referred to as food poisoning. The CDC estimates that *Salmonella* causes one million cases of foodborne illness in the US each year. Outbreaks have been linked to contaminated poultry and eggs, as well as cucumbers, pistachios, raw tuna, sprouts, and many other foods. Cattle are sources of *E. coli* 0157, resulting in contamination of hamburger and other meats that have to be recalled. Studies show that improperly cleaning the counter or cutting board used to prep the meat can spread the infection. Another common route of contamination is dipping a barbecue brush back into a bottle of sauce after brushing uncooked meat. We can also indirectly transfer potentially harmful microorganisms via our hands or contact with other surfaces into the nose, mouth, eyes, open cuts and wounds.

Lab studies show that bacteria and viruses that spread to environmental surfaces from an infected or carrier source can survive in significant numbers for several hours. In some cases they survive for days, especially on moist surfaces. Organisms can spread rapidly from a contaminated surface via hands, sponges, cleaning cloths, and hand and food contact surfaces around the home. Although raw food is a primary source of unwanted microbes in the kitchen, there is evidence that surfaces such as draining boards, sinks, dishcloths, and cleaning utensils are major reservoirs for bacteria. Frequently wet and moist sites, in particular, may act as permanent sources of bacterial populations.

MYTH: Cross-contamination does not occur in the refrigerator as it's too cold in there for microbes to survive.

FACT: Some bacteria can survive, and even thrive, in cool, moist environments like the refrigerator. For example, *Listeria monocytogenes*, a pathogen that contaminates meats and cheeses, loves to grow at 4°C (39°F). Your fridge's produce compartment may be one of the "germiest" places in the kitchen. Keep fresh fruits and vegetables separate from raw meat, poultry, seafood, and eggs. Clean the fridge on a regular basis with hot water and soap (this includes the walls and undersides of shelves) and clean up any spills immediately. Dry with a clean cloth towel and/or let removable parts dry in the air outside of the fridge.

Kitchen sponges are a hotbed of microbes. Researchers in Germany sequenced the microbial DNA of fourteen used kitchen sponges. They found 362 different species of bacteria living within them. Density was incredibly high: About 82 billion bacteria were living in just a cubic inch of space. In metric, a single cubic centimetre could be packed with more than 50 billion bacteria, which corresponds to about seven times the number of people inhabiting Earth. Such high bacterial densities are only found elsewhere in feces, the lead scientist told a reporter.

Sponges are the perfect environment for bacteria acquired via food, the skin, and other surfaces. They are the ideal incubator for microorganisms: warm, wet, and nutrient-rich. One microbe that particularly excelled among the fourteen sponges studied was *Moraxella osloensis.* This bacterium is widespread in nature and lives on the human skin. It is primarily responsible for the smell of dirty laundry, which may be one of the reasons that kitchen sponges can stink so badly. While its risk cannot be assessed from this study alone, it is important to be aware of as you scrub your dishes and counters with that same sponge day after day.

Surprisingly, the popular myth of boiling or microwaving sponges to kill bacteria did not work during the course of the study: Regularly sanitized sponges did *not* have fewer bacteria than uncleaned ones. In fact, sponges that were regularly sanitized by their owners teemed with a higher percentage of certain bacteria closely related to species that can cause infections in humans. We approach this finding cautiously, however, because the bacteria identified on the sponges were only a weak indicator for pathogenic potential of the identified bacteria. Further, the researchers were not aware of any cases where infections from these particular bacteria were reported from the domestic environments studied.

Given the high bacterial load of sponges, these scientists recommended replacing a kitchen sponge with a new one about once a week. To minimize cross-contamination, give sponges different jobs: one for cleaning the counter and one for the dishes. Definitely replace if it starts to stink, which is a sign of bacterial inhabitation. To reduce waste, the researchers suggested running it through a laundry machine at the hottest setting using detergent and bleach. Then use it elsewhere in the home in less hygiene-sensitive spots like the bathroom. Plenty of companies offer solutions such as bacteria-killing baths for sponges and antimicrobial countertops. However, without enough peer-reviewed scientific studies, it is difficult to evaluate their effectiveness. Plus, antimicrobial products pose risks as we will see later on.

From Cleanliness to . . . Super Bugs?

After relocating to Minnesota from Ontario, Jessica was horrified to learn that many Americans keep their shoes on *inside* the house (generally not a Canadian custom). She cringed at the thought of all the dirt, salt, and grime trailing through the home, especially on her freshly scrubbed floors. However uncomfortable it may be, while researching for this book Jessica learned that in general it would better if more people brought the outdoors inside—at least when the weather is fair and mud/snow/salt is kept to a minimum. She adjusted some of her everyday habits to find a more holistic approach to "clean."

Many microbiome researchers reason that people are doing themselves a disservice by waging a war on germs at home. Any approach that sterilizes your home gets rid of both the bad *and* good bacteria, with potentially harmful consequences. It leads to a critical question: To what extent do we need to stop protecting people from germs, and instead protect germs from people?

An antimicrobial is designed to kill or inhibit the growth of microorganisms including bacteria (antibacterials), viruses (antivirals), and fungi (antifungals). Until relatively recently, the primary kinds of at-home antimicrobial products were disinfectants, antiseptics, and antibiotics. Disinfectants are products that kill microbes on surfaces like countertops or toilet seats. Antiseptics are products designed for use on skin issues like cuts and scrapes. Antibiotics, as we know, are intended to destroy microbes within the body. Now a new class of products is emerging as companies race to put antimicrobial ingredients into all sorts of items.

You can find antimicrobial products in homes, public spaces, workplaces, and schools; in clothes, cutting boards, and furniture. The US Environmental Protection Agency (EPA) regulates antimicrobial products as pesticides (disinfectants), while the FDA regulates antimicrobial products as drugs (antiseptics and antibiotics). As pesticides, antimicrobial products are used in everyday objects such as countertops, toys, grocery carts, clothing, utensils, and hospital equipment. As drugs, antimicrobial products are used to treat or prevent diseases on and within people, pets, and other living things. Think of the distinction between these domains as the difference between wipes for the kitchen or bathroom (regulated by the EPA) versus hand-sanitizing wipes (regulated by the FDA). EPA registration is helpful because it ensures that the product actually does what it claims to do, if used properly. Clorox Disinfecting Wipes, for example, require enough wipes to be used so that surfaces remain visibly wet for four whole minutes to disinfect (i.e., kill the bacteria, viruses, and fungi on surfaces or objects), and at least ten seconds to sanitize (i.e., lower the number of microbes). Personal care products like hand soaps, toothpaste, and deodorants are regulated fairly loosely by the FDA. They are more focused on antibiotics and other drugs, which are fastidiously regulated to assure their safety, effectiveness, and security.

The marketing of antimicrobial products preys upon consumer fears, such as frightening advertisements with stories of *E. coli* outbreaks, scary-looking viruses, and drug-resistant germs. They claim to "kill 99.9 percent of germs." Do these products work? Does overuse of antimicrobial ingredients help breed super-germs? It is a controversial topic in the public health realm today. The US government recognizes antimicrobial resistance as a growing global threat; at least two million Americans every year are infected with antibiotic-resistant bacteria, and at least twenty-three thousand people die each year as a direct result of these infections.

There is ample scientific evidence to suggest that ubiquitous and non-judicious use of antimicrobials has negative consequences. Many products include the ingredient triclosan, for example, which even in concentrations of 0.2 to 2 percent have antimicrobial activity. Triclosan kills bacteria but has little to no effect on viruses. This is important because most common household illnesses like colds and the flu are viral, so antibacterial ingredients like triclosan will not prevent them from spreading.

Public health experts fear that overly heavy use of antimicrobials, especially in the uncontrolled home environment, may result in microbes resistant to these chemicals. Triclosan may cause a resistance to develop because its mechanism of action is so very specific and its use is becoming so widespread. In fact, resistance to triclosan has already been observed in laboratory studies. Given this, many experts recommend avoiding household products that contain triclosan. There is minimal evidence that they are effective in homes, and considerable concern about growing resistance. This information is pertinent to independent and assisted living settings where antimicrobial applications might be over-used. These everyday products could do more harm than good for aging residents.

Instead of making our living environments as sterile as possible, to support healthy aging, we need to strive for the right kind of hygienic clean. First of all, a very small fraction of the microbes that we encounter in the environment are hazardous. In fact, as we have seen in this chapter, many are beneficial. There is no such thing as a microbe-free home. Disinfectants kill microbes on surfaces temporarily, but they do not provide long-lasting disinfection. In some situations disinfectants are needed, but in many other occasions they are not. Disinfecting a toilet bowl, for example, is an exercise in futility. While it is important to regularly clean the toilet, given its regular exposure to fecal matter, sterilization is impossible. Another example is spraying a room deodorizer disinfectant. One cannot disinfect air this way. If something smells funny, find the source and clean it up. When might you really need a disinfectant? If the sewer overflows in your basement. Or if highly susceptible individuals with special health problems live in a home. For most other occasions at home, simply cleaning with soap or detergent and clean water should be adequate as long as it is done frequently and thoroughly. Simple cleaning and handwashing practices generally remain one of the most effective ways to prevent the spread of many types of infections and illnesses.

Housekeeping, Microbe Style

In this chapter, we saw that a very small fraction of the microbes we are exposed to are as hazardous as we think—and many are actually beneficial. Microbes are constantly all around us, even in interior, supposedly "clean," environments. The indoor microbiome is affected by factors such as air currents and ventilation, surface contact, plants, and human and animal interactions. As we are better able to identify the functions of helpful indoor microbes, we can better plan for building design, operation, and function to maintain the health of people who live and work in these spaces by promoting the right kind of microbe exposure. Instead of nursing homes, hospitals, and care environments that are as "sterile" as possible, we will design environments that are comfortable living spaces for us and our microbial friends. We will work with them to achieve longer and healthier lives.

Current cleaning and hygiene standards in institutional settings, especially hospitals and nursing homes, tend to promote multi-resistant pathogens instead of supporting beneficial microbes. We're beginning to see important steps taken to address mounting concerns regarding the insertion of antimicrobial chemicals into everyday products. This includes the FDA's ban on soaps containing nineteen different antibacterial chemicals (including triclosan and triclocarban). Moving forward, it is essential to rethink our understanding of both sterility and our relationships to surrounding microbes. The next generation of microbial sequencing and microscopic techniques will enable us to better assess indoor microbiomes so that we can develop management strategies for beneficial interactions. Age-friendly environments should foster the health and well-being of all inhabitants—microbes included.

For now, removing bad microbes at home requires few interventions. Ventilate with outdoor air whenever possible, keep surfaces clean, and reduce heavy carpeting and other flooring that traps dust microbes. Indoor houseplants and flowers can positively influence health by introducing beneficial bacteria indoors, in addition to improving air quality. Further increase microbial diversity in the home with pets, regular visits from friends and family, and cleaning with soap and water instead of antimicrobial products. It may even be beneficial to keep your shoes *on* in the house (just watch where you step!). Do not confine yourself to a climate-controlled, sterilized environment as you age.

QUICK TIPS

- **The great outdoors:** Playing in the yard is not just for kids. Adults increasingly quarantine themselves in sterile, climate-controlled indoor environments, which has detrimental effects on well-being. It is important to get outside at any age. Also find ways to bring the outside in, such as household plants and fresh air.
- **Put the phone down:** Microbes are another reason to divorce yourself from your cell phone. Leave it behind on your next trip to the bathroom and at the kitchen table. Clean it regularly to minimize the number of potentially harmful microbes pressed against your face.
- **Kitchen culprits:** To avoid cross-contamination, scrub kitchen surfaces frequently with soap and water. This includes cutting boards, counters, microwaves, and refrigerators. Sponges are a hotbed of microbes and therefore need to be replaced on a regular basis—ideally once a week.

- **The right kind of clean:** Avoid products with antimicrobial ingredients, especially triclosan. Plain old soap will do the job in nearly all instances. Even after handling raw chicken, for example, washing your hands thoroughly with soap and water is completely adequate to remove unwanted bacteria. Same with the surfaces in your home; simply clean with soap or detergent, water, and a rag/towel for everyday purposes. The physical action of scrubbing and then rinsing with water removes substantial numbers of organisms.
- **Handle hygiene with care:** Institutional care centres such as nursing homes and assisted living facilities should try to cultivate as microbe-friendly an environment as possible. Open the windows, nurture houseplants, and move in with pets (where allowed). Do your homework beforehand by asking about the institution's cleaning practices, history of infectious outbreaks (such as gastrointestinal and respiratory disease), types of foods served, and scheduled activities. Check out the facility in person and try to speak with current residents and staff about their conditions. Find out what a typical day looks like. How does the food look and taste? What sounds and smells do you pick up on?

14

The Fountain of Youth *Is* Full of Microbes: The Future of the Whole-Body Microbiome

Back in Chapter 11 we met Élie Metchnikoff, the father of immunology, who over a hundred years ago advocated consumption of lactic cultures (containing beneficial microbes) to preserve health and longevity. He observed that Bulgarians living in the countryside lived longer than richer Europeans, despite suffering from poverty and a harsher climate. Observing that countryside residents consumed fermented milk, Metchnikoff concluded that the lactic acid bacilli that ferment the milk had anti-aging benefits, and therefore we should consume fermented milk for longevity. He published this work in a now-famous book, *The Prolongation of Life*, and won the 1908 Nobel Prize in Physiology or Medicine for groundbreaking work in immunology. Metchnikoff believed that there were "good" and "bad" microbes in the gut. The "bad" ones produced toxins as part of their metabolism that caused damage to the human body—he called it "intestinal autointoxication"—while the "good" ones produced beneficial fermentation products, such as lactic acid, that could counter harmful products and promote health. This was the birth of probiotics and the concept that gut microbes could be manipulated for health benefits and longevity.

Metchnikoff's hypothesis is eerily prescient of what we are just "learning" today—over a century later—about the role of microbes in our short- and long-term health. Reinterpreting his thoughts, we now know that the harmful toxins are inflammation-inducing microbes and their products, while the "good" ones are those that produce anti-inflammatories such as SCFA. The idea that we might be able to manipulate microbes and their products to benefit health and aging remains a very exciting one, and great strides in scientific and clinical applications are on the horizon. The microbiome is profoundly affecting fields of health and medicine across all areas of study of the body, providing new diagnostic techniques and therapies for a host of conditions that couldn't have been imagined a few short years ago. However, even our most "advanced" current therapies used in the clinic, such as fecal transfers, are still very crude. They will require extensive research to fully take advantage of the microbiome, and the various molecules that microbes make that have an effect on our bodies. This final chapter will take a critical and in-depth look at existing methods used to manipulate the microbiome such as diet, probiotics, and prebiotics. The applied field of microbiome research is full of much hype and hope, and many false or exaggerated claims that are not yet based on sound science. Here, we try to provide some guidance based on the current scientific knowledge regarding what you can do today to enhance healthy microbes and their benefits for well-being at any age. We also include a section on what to discuss with your physician based on individual health needs and concerns—or in the case that you are a caregiver, on behalf of the care recipient. At the end, we peer into the future to suggest novel concepts that are on the horizon for this rapidly advancing field.

You Are What You Eat

"Eat your vegetables; they are good for you!" Every child has heard this hundreds of times. However, parents should really say, "Eat your vegetables; they are good for your *microbes*, which are good for you." Diet has a major effect, good and bad, on our microbes. By far the easiest way to improve your microbiome's composition is to change your diet. Despite many undocumented claims, the exact details of a "healthy" microbiota and the type of diet needed to achieve it are not fully understood—and as Jessica saw firsthand in her PhD research, "healthy" won't mean the same thing for every person. However, several guiding principles regarding diet and how foods affect and shape microbiota are becoming well established. Following these principles may be particularly important to older adults given the significant changes that occur in the microbiome later in life.

A study following older people who transitioned from living independently in the general community to living in long-stay residences found that microbiota changes do not precede, but actually follow, shifts in health status, possibly due to dietary changes. 92 percent of those in long-stay residences consumed a high-fibre/low-fat diet (which is associated with having lower microbial diversity), while only 2 percent of the general community consumed a similar diet (the remainder followed a higher-fat/low-fibre diet). A high-fibre/low-fat diet showed higher levels of health-promoting SCFA such as butyrate. Within one month of being transferred to a long-stay residence, everyone adopted a long-stay diet; yet it took a year for their microbiota to be clearly differentiated from those of the community dwellers. Researchers concluded that microbiota changes, which were probably mediated by dietary shifts, occur *after* changes associated with aging that put the subjects of this study in the residence in the first place.

Besides fat, fibre was a key dietary factor in the study above. As we discussed in Chapter 7, it is broken down into SCFA by gut microbes. One study showed that supplementing a diet with fibre increased microbial diversity up to 25 percent in those who started with low diversity in their microbiota. Resistant starches (a type of fibre) have been shown to affect the microbiota by improving metabolic measures (such as decreasing glucose spikes) and lessening inflammation. Unfortunately, the typical Western diet has very little fibre in it, while our ancestors ate much more fibre. This change not only affects us today, but there is even concern about long-term changes in the composition of the entire species's microbiome. Mouse studies have shown that while consuming a low-fibre diet has a reversible effect on the microbiome in a single generation, after multiple generations it is impossible to revert back to a high-fibre microbiome without also adding back the missing microbes (see Chapter 7). This suggests that if humans continue low-fibre diets over the generations, beneficial fibre-consuming microbes could go extinct. Current recommendations suggest consuming at least twenty-one to twenty-five grams of fibre per day for women, and thirty to thirty-eight grams a day for men. High-fibre foods include raspberries, pears (with skin), whole-wheat spaghetti, barley, bran, split peas, lentils, beans, artichokes, and green peas.

Again, an easy place to look for inspiration on this path are the Mediterranean and MIND diets—which are rich in fibre, vegetables, fruits, nuts, and legumes, and suggest decreased amounts of dairy fats and red meat. The MIND diet has been associated with health boosts in multiple studies, including decreased risk of Alzheimer's disease by up to 53 percent. Another study looked at 1,296 aging individuals from five European countries. The experimental group was given dietary instructions and vitamin supplements and asked to follow a Mediterranean diet. Compared to controls, those following the diet instructions lost less bone mass and showed reduced inflammation compared to the controls. Included in the high-fibre diet are whole grains, which, according to studies, appear beneficial to humans on a Western diet. In just a six-week period, those who followed a whole-grain diet saw enhanced beneficial microbiota by increasing SCFA producers, decreasing inflammatory bacteria and inflammation, and enhancing weight loss.

In part produced by microbes, fermented foods are also a good way to increase beneficial microbes travelling through your gut (think of Metchnikoff and his fermented milk). There is little doubt that these probiotic foods, such as sauerkraut and kimchi, provide additional nutritional value, but we lack strong clinical research to confirm their potential health benefits beyond nutrition. Few controlled human trials have been carried out with fermented foods other than probiotic yogurts. This handful of small studies showed improvements in weight maintenance and reduction in cardiovascular disease, type 2 diabetes, and metabolic syndrome. Although these foods seem to have beneficial effects, we still cannot say with certainty that they are working through the microbiome.

What has complicated studies about diet and microbiota is the difference in microbiomes between individuals. As we saw in Chapter 7, individuals respond differently to various food groups. This suggests that personalized diets based on the microbiome, such as those offered by DayTwo, could become a mainstay of diets in the future. It will also allow us to establish the effect of different microbes on food groups, and the effects of different foods on the microbes—this information is really needed if we are going to truly understand the role of the microbiome in diets and nutrition.

No matter what nutritional or diet program you follow, aim for stability and continuity. We saw the harms of yo-yo diets in Chapter 7 and of travel disruptions in Chapter 13. If you go on and off diets, there seems to be a microbial memory that sabotages weight loss efforts. Remember, also, what we saw with artificial sweeteners. In a subset of people, they may actually cause dysbiosis and worsen metabolic symptoms by increasing glucose intolerance.

Caloric restriction has been shown to increase longevity in many organisms including rats, mice, fish, flies, worms, and yeast. However, because purposefully testing severe calorie restriction on human beings is unethical, its effects in humans remain unknown. Exactly how caloric restriction works in animals is not yet known despite over seventy years of research, although countless theories abound. In a mouse study, animals subjected to lifelong calorie restriction had significant changes in their microbiome. They generally had increased numbers of microbes associated with health and decreased numbers of those associated with inflammation. Those on low-fat diets had very high levels of *Lactobacillus*, which was strongly correlated with longevity.

For anyone contemplating going on a calorie-restricted diet per the terms of the study, note that it's not your typical "diet." You would have to decrease calories by about 30 percent... for the rest of your life. For the average 2,000-calorie diet, that means cutting down to 1,400 calories a day! You might live longer, but you would be in a perpetual state of starvation—hence the difficulty of testing in humans.

While the exact correlations between food groups and microbiota are still lacking, the general concept of increasing fibre to benefit your microbes and decrease inflammation seems to be a salient one.

> **Recommendation:** *Pursue a diet that is rich in fibre, including fruits, nuts, vegetables, whole grains, berries, fish, and moderate amounts of red wine. Avoid red meats and foods rich in animal fat. Diets such as the Mediterranean and MIND diets seem to be generally good ones to follow from both a healthy human and microbial point of view.*

Probiotics for Life

There is no topic that generates more discussion in the microbiome world than probiotics. It is a gigantic industry estimated at over thirty billion dollars per year worldwide. While products such as kombucha have risen rapidly in popularity, a bigger question remains: Do they work? A simple question with a complex answer. In this section, we will dig deep into probiotics so that you can have a more informed opinion on this growing trend and how it might help you.

Probiotic is a term that originates from Greek and means "for life." Using probiotics for health has a long history: In 1917, German scientist Alfred Nissle isolated a non-pathogen *E. coli* strain from the feces of a World War I soldier who did not contract an infectious diarrhea (shigellosis) that was rampant at the time. This bacterial strain became known as *E. coli* Nissile 1917, which is now an accepted probiotic. The current definition of a probiotic, according to WHO and other organizations, is "live organisms that, when administered in adequate amounts, confer a health benefit on the host." Probiotics are live microbes, usually species of *Lactobacillus*, *Bifidobacterium*, some *E. coli* and *Bacillus*, and the yeast *Saccharomyces boulardii*, that, when taken regularly in sufficient doses, are designed to promote health effects. They can be taken separately or added to food. The problem is that there are many probiotics now available but not much clarity or consensus about what each of them is, what they do, and who should take them. When the average consumer walks into a health food store, they are confronted with a bewildering array of probiotic choices—usually one or more walls covered in shelves, each with its own enticing but unchecked health claims.

Another problem is that despite a large body of literature on probiotics, we still do not really know how they work. They seem to have a variety of effects, including decreasing gut permeability and decreasing inflammation (which are likely interrelated). They do not seem to have a direct long-term effect on the overall structure of the existing microbiota population given that they rapidly disappear from the gut. Recommended doses are so large—one hundred million to ten billion or more live microbes per day—because they are largely isolated from the environment, animals, or other body sites. They aren't meant to live in the gut, so they don't colonize there.

The third issue is that because probiotics are currently sold in Canada and the United States as foods and dietary supplements, they are not regulated by the FDA. Their health claims are often, then, not supported by the kinds of extensive and rigorous clinical trials required for FDA-approved drugs. Although there are over 1,500 clinical trials involving many different probiotics, they are often not well-designed studies: Many are not randomized, and they are difficult to conduct completely blinded (participants often know or can figure out if they are in the control or study group). There is little good clinical data on the ability of probiotics to promote overall health (as opposed to helping heal a specific disease). Still, many people swear that probiotics make them feel better. Probiotics rarely have any side effects and in general all are considered safe; they are also generally quite cheap compared to most prescription drugs one could use for similar conditions.

The fourth issue is that most probiotics indicate how many live bacteria are *supposed* to be in the supplement, but this does not always reflect the actual viable counts. Factors such as refrigeration and other storage techniques can affect viability. The lack of standardization in the field causes large heterogeneity in the quality of different probiotics. There are significant issues with labelling: One study found that a product did not even contain the supposed *Lactobacillus* on the label, and only 80 percent had the viability that the labels claimed. They are still generally safe to take; they just may be less effective than advertised.

So how does one make sense of all the confusing and often conflicting information out there about probiotics? There are groups dedicated to aligning probiotic information and claims with actual science. One such group is the International Scientific Association for Probiotics and Prebiotics (ISAPP). Dr. Mary Ellen Sanders is the founding president and current executive science officer of ISAPP, and got in at the "ground floor" of probiotics in the United States when she entered the field in 1990. "I was at first a big skeptic. In my graduate school days, I thought that the whole probiotic area was just snake oil. To this day, my graduate advisor teases me that ever since I went out on my own to start a consulting business, probiotics have consumed everything that I work on. My skepticism was not unfounded. At first it was much more a field of faith than science." Now, after decades of serious research and advancements in high-quality human trials, Dr. Sanders feels confident saying that probiotics have important effects on health. She believes that probiotics will flourish if the available products are scientifically validated, responsibly produced, and accurately labelled.

As a steward of the field, Dr. Sanders, along with prebiotic expert Professor Glenn Gibson, formed ISAPP back in 2002 to connect multidisciplinary scientists. This involved researchers from microbiology, biochemistry, gastroenterology, and food technology, as well as the wide range of people more directly involved in probiotics and prebiotics.

With the aim of bringing great minds together, ISAPP has also begun to focus on the abundance of information—and misinformation—out there. The organization offers helpful information on what to look for in a probiotic and how they might work. Jargon-free infographics, a few videos, and detailed information are easily accessible to the general public at their website (isappscience.org).

In addition, two excellent websites—Clinical Guide to Probiotic Products Available in USA (usprobioticguide.com) and the Canadian equivalent (probioticchart.ca)—are designed to give a synopsis of the clinical evidence regarding a particular probiotic and specific conditions. They are updated yearly and provide a wealth of accurate information about which probiotics may work for which condition based on clinical evidence distilled from the numerous papers describing studies of various probiotics. Although the sites are designed for clinicians, they are terrific tools for everyone wanting to understand which probiotics have actual clinical data and, just as important, how strong that data is. If the websites appear confusing at first glance, it is well worth the effort to work through them and study the guides to determine what probiotics may work, which ones to use, what doses, and the underlying clinical evidence.

Both websites include a series of tables for adult health, pediatric health, vaginal health, and functional foods with added probiotics. Each table lists many of the commercially available probiotics in either Canada or the US by their name, which probiotic strain(s) they contain, the dose, and the number of doses needed per day. What follows is a ranking of the clinical evidence available for a series of diseases. A ranking of I is the highest, meaning there was efficacy seen in at least one well-designed randomized clinical trial. Levels II and III have less clinical evidence available. Next is a list of diseases, with each disease abbreviated by one or more letters (e.g., ID = infectious diarrhea; AAD = antibiotic-associated diarrhea). Hover the mouse cursor over the abbreviation, and the actual disease name will pop up (the abbreviations are also listed at the beginning of the site).

There are a few examples where we do have convincing clinical evidence of probiotics' efficacy in treating specific conditions. These include antibiotic-associated diarrhea, irritable bowel syndrome (IBS) and associated depression, prevention of *Clostridium difficile*-associated diarrhea, adjunct therapy with *Helicobacter pylori* eradication, colic for breastfed infants, vulvovaginal candidiasis, and bacterial vaginosis. The bottom line is that some probiotics can work for some conditions, although many do not show clinical efficacy in spite of extensive marketing and sales. All these uncertainties about probiotics will hopefully change as scientists rapidly investigate specific mechanisms of the human microbiome and effects of probiotics. Armed with this knowledge, Dr. Sanders expressed excitement regarding the continuing evolution of probiotics to meaningfully intervene in health and disease.

> ***Recommendation:*** *Some probiotics may work well for conditions including antibiotic-associated diarrhea,* C. difficile*-associated diarrhea, IBS, colic in breastfed infants, and bacterial vaginosis. However, there are many others that lack proven ability. Take the time to consult the websites usprobioticguide.com and probioticchart.ca and talk to an expert before incorporating probiotics into a specific healthcare regimen.*

Probiotics 2.0

Given the major recent advances in our understanding of the microbiome, there is strong enthusiasm for a future of next-gen probiotics—what we are calling probiotics 2.0. These will be microbes isolated for use in a specific body part, such as using gut microbes for gut disorders or vaginal microbes for gynecological infections. They may be a combination of microbes rather than a single microbe; they may be taken in small doses, since they are designed to live specifically in that one body site, and thus won't be rejected as a foreign invader; they may readily colonize that body site; and most importantly, they will have a defined beneficial function.

A subsection of this new generation of probiotics are being developed and approved by the FDA as drugs, known as live biotherapeutic products (LBPs) or live biotherapeutic agents (LBAs). In Chapter 11, we discussed a mixture of *Clostridium* species that had a strong effect on regulatory T cells. Brett is involved with Vedanta Biosciences, Inc., a biotechnology company in Cambridge, Massachusetts. They are about to enter clinical trials using this mix of microbes for a variety of applications, including infections associated with *C. difficile* or multiple resistant organisms. Another initiative Brett is involved in is Commense, also based in Cambridge. They are looking to provide infants a defined cocktail of microbes early in life (three months) in an attempt to decrease the risk of asthma later in life. They also hope to transmit vaginal microbes to children delivered by C-section in order to decrease the risk of asthma, allergies, and obesity normally associated with C-section deliveries. (Brett is a scientific co-founder of both these companies.) Elsewhere, there is significant interest in using *Faecalibacterium prausnitzii* and *Akkermansia* to increase SCFA production in the gut to decrease inflammation. Many companies are also working on the use of novel bacterial strains as probiotics, based on new knowledge gleaned from studies on the microbiome. Another area of intense interest (but without hard data yet) is "psychobiotics"—probiotics that might affect the brain to treat conditions such as depression and anxiety.

Yet another frontier is genetically engineering bacteria. Work underway includes designing probiotics that express anti-inflammatory cytokines (to directly affect inflammation), or useful metabolic pathways such as producing SCFA including butyrate, which will hopefully have beneficial health effects. The sky is the limit in designing synthetic microbes, although there are massive regulatory approval hurdles to overcome because this involves releasing a live genetically engineered microbe into the environment. For example, *E. coli* Nissle 1917 has been engineered to inhibit the virulence of *Vibrio cholerae* by disrupting how the vibrio senses its environment. It worked well in a mouse cholera model but has not been tested in humans. Although most of the genetic engineering has been done in approved probiotic strains (*E. coli* Nissle 1917, lactobacilli, and *Bacteroides* species), using novel microbiota strains in combination with genetic modifications might enhance delivery of certain compounds such as anti-inflammatories, as they would colonize body sites more efficiently than current probiotics. Like all new technologies, there are concerns with this approach that need to be seriously addressed. For example, *Akkermansia* has been shown to be extremely beneficial in many gut-associated diseases, yet it is also associated with increased risk of disease in several neurological disorders such as Parkinson's, dementia, Alzheimer's, and multiple sclerosis. This suggests that serious safety concerns need to be addressed before accepting this as a widely used probiotic.

The other concern for any new probiotic is that it would be introduced into the environment and could have unforeseen consequences. The advantage of current probiotics is that they do not stick around long; however, new microbes that colonize well and propagate have the potential to spread into the environment (including other people) with unknown effects. "Kill switches" can be genetically introduced into microbes and have been used in live engineered vaccine strains; they may be necessary to incorporate into new probiotics.

Recommendation: *Stay tuned; exciting new designer probiotics are coming to a store near you!*

Prebiotics

Instead of taking probiotics, which supplement an existing microbiome, how about eating certain foods that enhance the growth or activity of your beneficial microbes from the get-go? These are called *pre*-biotics. The definition of a prebiotic has evolved over the years, but it generally refers to a substance that is selectively fermented (broken down) and results in a specific change in the composition and/or activity of the gastrointestinal microbiota. This in turn confers benefits upon the host's health. Thus, the food itself is not broken down by the host (it is non-digestible) but instead the bacteria metabolize it, with beneficial effects on the microbiome as well as the host.

Prebiotics are a subset of fibre (carbohydrates or sugars, often called resistant starch) that includes inulin, fructooligosaccharides, and galactooligosaccharides. These microbial foods selectively enrich the beneficial *Lactobacillus* and/or *Bifidobacterium* species (although we now realize that other microbes are also affected by these prebiotics, especially buty-rate producers). They are thought to also increase microbial diversity. Prebiotics are naturally found in plant foods and are usually isolated from biological sources.

There are many hypotheses about how prebiotics work. The theory is that by consuming foods that in turn feed these beneficial microbes, the microbes will increase in number and provide benefits such as increased SCFA production. They could also affect how long it takes materials to move through the intestinal tract (known as gut transit time), viscosity (thickness of the gut material), and how microbes interact with other food components.

Like probiotics, the most pressing question is: Do they actually confer a health benefit to the user? Again, the data are not well established and require more studies. For some, they definitely enhance the number of beneficial microbes, but actual proof of providing health benefits is not yet clinically established. Some studies have also shown that inulin and fructooligosaccharide enhance calcium absorption and bone density, especially in teenagers, while contrasting results were obtained in adults. There are reports that they may affect vaccine responses, as well as have beneficial effects for infectious diarrhea and IBS. In studying the collective literature on prebiotics, the general consensus is that they may enhance some health effects, but the data are not as strong as for probiotics. Additional confirmatory clinical studies are needed.

In the meantime, Brett regularly sprinkles a tablespoon of inulin on his morning (whole grain) cereal. Although he can't swear to a health benefit, it is a fun experiment as he then gets a real gut-check and can appreciate first-hand the gut microbes' fermentation in action (think:

gas!). Prebiotic foods high in plant fibre include raw chicory root, Jerusalem artichoke, garlic, leeks, onion, and asparagus.

Overall the concept of prebiotics makes sense: Consume substances that select for beneficial microbes and sit back and let them do their job. However, translating this into reality has been more difficult than it seems. Similar to probiotics, as we learn more about microbial metabolism in the gut, better prebiotics will probably be developed that will have health benefits through defined mechanisms.

> ***Recommendation:*** *There are too few appropriate human intervention trials with prebiotics to definitively say that prebiotics work in people and have a beneficial effect, although several more clinical studies are underway. At present, eating a balanced diet rich in a variety of different fibres has a likely beneficial effect on the microbiome, and can contribute to improved overall health. There are currently no reasons not to use them as they have few side effects (other than dose-dependent flatulence as a by-product of fermentation).*

Synbiotics

If there are strong indications that probiotics and prebiotics each might have beneficial effects, what happens if you combine them? These are called synbiotics. In theory it's a great idea, but there are no data yet to say that they work better as a team than prebiotics or probiotics alone. In one study, older adults (aged sixty-five to ninety) were given the probiotic *Bifidobacterium longum* and the prebiotic inulin. This treatment did increase bifidobacteria numbers (as well as other beneficial microbes) and decreased Proteobacteria (pro-inflammatory). They also saw an increase in butyrate production and a decrease in pro-inflammatory cytokines; however, these beneficial effects soon disappeared after the treatment ended. As great as this increase sounds, even in isolation, there are many other studies that did not see any major effects from doses of synbiotics or found that individually a probiotic or prebiotic was more effective than the combination.

Recommendation: *There is not enough good clinical data yet to recommend the use of synbiotics, although both probiotics and prebiotics on their own can be beneficial. We just don't know yet if their effects are synergistic.*

Antibiotics

Antibiotics are one of the biggest medical discoveries in the twentieth century and have saved countless lives. They are the biggest reason we have a longer lifespan than people did a century ago. However, antibiotics are also a source of new problems, including how they affect the microbiome. Over 80 percent of antibiotics in Canada and the US are not used to treat disease but rather in animal husbandry as growth enhancers; in Europe this practice is banned. North America is finally starting to take notice, with huge companies such as McDonald's, Subway, and Kentucky Fried Chicken now using antibiotic-free livestock and poultry. Use of antibiotics in animal husbandry, in addition to over-prescription by physicians for conditions that are not improved by antibiotics, and abuse of antibiotics by the public, has led to antibiotic-resistant pathogens. These "superbugs" are very difficult to treat and cause legitimate concerns about a global return to a pre-antibiotic era. As we saw earlier, at least two million Americans are infected annually with antibiotic-resistant bacteria, and at least 23,000 people die from these infections each year in the United States.

This may have you wondering: Is it safe to use antibiotics at all? Without hesitation, yes—but only if they are warranted! If you face a life-threatening infection, they can save your life. However, if you have an ear infection that may be viral (in which case antibiotics won't work anyway), physicians will often suggest waiting a day or two before throwing a microbial carpet bomb on your system with a round of antibiotics. As we saw in earlier chapters, antibiotic use is associated with obesity, asthma, allergies, stress, depression, and many more issues. It really comes down to balancing their use and weighing the risks. If you need to take antibiotics, we have discussed various ways of potentially repairing your microbiome afterward, such as certain probiotics that work against antibiotic-associated diarrhea, as well as a varied diet that enriches for microbial diversity and beneficial microbes.

> **Recommendation:** *If there is an evidence-based medical recommendation that antibiotics are warranted (such as a serious bacterial infection), take them. But remember that antibiotics can have long-term effects and don't work against viral infections. If they are not warranted, don't use them. If you do need antibiotics, consider diet and probiotics to try and repair your microbiome after the treatment is finished.*

RePOOPulation: Fecal Transfers

In all our discussion of fecal microbiota transfers (FMTs), there's a clear bottom line: They can work very well for infections with *C. difficile* and show significant promise for ulcerative colitis (although it can be donor dependent). Additionally, there is real promise being seen for fecal transfer use with stem cell transplantations, preventing multi-drug resistant infections such as Vancomycin-Resistant Enterococcus (VRE) and multi-drug resistant Enterobacteriaceae, and possibly for use with checkpoint inhibitor immunotherapy in cancer. For most uses, they are currently experimental and being tested for a variety of conditions.

These advances are exciting, but again, we must stress: Do NOT try this at home. If the bowel is perforated during an attempted transfer, the individual could die of sepsis. There are also no studies yet on long-term risks associated with FMTs, including the concern that an undesirable phenotype may be transferred from the donor feces. Remember the example in Chapter 7 of the lean person who gained significant weight after receiving an FMT from an obese donor. Although the FMT successfully cured the patient, it may also have shifted the recipient's microbes towards a pattern of obesity.

In the near future, we will be able to use a defined batch of fecal microbes grown in the lab—a process affectionately known as rePOOP-ulation amongst microbiologists (see Chapter 7). This will eliminate the risk of spreading unknown infectious agents found in feces (which is a body fluid). The use of synthetic communities, or live biotherapeutic agents, as discussed above, is rapidly being developed by various companies to replace donor FMTs in the future.

> **Recommendation:** *If you have had at least two standard antibiotic treatments for recurrent* C. difficile *infection without persistent cure, ask your doctor about a fecal transfer. The clinical data to support its efficacy is strong. If you have ulcerative colitis, you could try to get into a clinical trial where FMTs are being used. For all other indications, clinical evidence of efficacy is still lacking for FMT treatment, so it is not warranted. Do NOT try an FMT at home.*

Talk to Your Doctor

Playing an active role in your healthcare is one of the best ways to prevent and treat disease, since you know your body best. Open and honest patient-physician communication is essential to conveying the right information at the right time so you can receive the best and most appropriate care. Medicine is not a one-way relationship where the doctor takes the lead and you (the patient) follow without questioning. Rather, think of yourselves as partners, and the two of you as part of an even larger team including other physicians, nurses, pharmacists, therapists, assistants, and healthcare providers, all necessary to keeping you healthy. Be honest and upfront about your symptoms and opinions, and don't be afraid to ask for other opinions. If you are meeting with a specialist or have a critical appointment coming up (beyond a routine check-up or exam), here are a few general tips to help make the most of your visit:

1. Write down a list of questions and concerns before the appointment. If you consulted medical literature in advance, print out a copy and bring any notes with you.
2. Bring a list of all prescription drugs, over-the-counter medicines, vitamins, probiotics, prebiotics, and any other supplements that you take.
3. Consider bringing a trusted family member or close friend with you to assist with moral support and clear communication lines with the doctor. Let your companion know in advance what you want to get out of your visit to help you stay on track. He/she can also assist in remembering details of the conversation.
4. Take notes about what your doctor says (or if you have a companion, ask him or her to do so).
5. Ask questions if the doctor's explanations or instructions are unclear. Bring up any problems or concerns even if the doctor does not ask if you have questions.
6. Clarify if an antibiotic treatment or any other procedure is necessary. Discuss potential risks clearly.
7. Learn how to access your medical records so that you can track test results, diagnoses, treatment plans, and medications.

These tips can be especially important as we age. There are often more health conditions and treatments to discuss, and it can be a period when health concerns have a bigger impact on quality of life and longevity, making such conversations more detailed and possibly stressful.

In talking to a doctor about your microbiome, be upfront about what microbe-friendly practices you are following and/or want to try. Remember, physicians are sworn to the Hippocratic Oath, which means first and foremost, do no harm. They need clinical proof that what they recommend will not harm the patient and will have a proven health benefit. For example, if you have experienced multiple bouts of C. diff and want to pursue an FMT, ask your healthcare providers about safety and effectiveness. Be as informed as possible regarding what the clinical data supports versus what is promising but "not just there yet." We spoke to

Dr. David Patrick, a practising infectious disease doctor, professor in the University of British Columbia School of Population and Public Health, and epidemiologist for the BC Centre for Disease Control. Dr. Patrick recommended that in considering any new microbial approaches to your health, science is the first place to start: "In the history of the medical profession, we've done a lot based on impulse without the science to back it up. We first need to prove things in experiments (in humans)—meaning rigorous clinical trials—before any widespread application of a treatment can be considered. The risk is that people will jump all over the coolness of microbial advances and leap to conclusions." He clarified that even though studies may show exciting results in mice or initial small human studies, that does not mean they will necessarily translate into successful results for all. For all new treatments, many expensive and long-term clinical trials need to be done before anything is adopted into mainstream medicine. The microbiome is just too new a field for most practices to be fully clinically approved.

Furthermore, shifting your microbiome will not only affect you. It will also affect those around you. As we saw in Chapter 13, we share and swap immense numbers of microbes with our roommates—human and non-human. In the worst-case scenario, an ill-informed microbial experiment can harm not only you but those you live with as well. That being said, "do no harm" does not mean "do nothing." Dr. Patrick explained that the first thing a doctor should do is show compassion regarding a patient's symptoms. With broad and ambiguous illnesses such as chronic exhaustion, there may be a need to turn to new treatments whenever the risk is low and science can reasonably back it up. Here the placebo effect may be able to help. Dr. Patrick explained: "Placebos can actually do a lot to make people feel better. It's not the sugar pill or saline that works, but rather setting expectations. Having the right expectations can meaningfully aid one's recovery."

Regarding antibiotics, Dr. Patrick told us that his standard response to patients is "only take them when they are needed to resolve a serious disease process." Similar to our recommendations, he recognized that there is a proper time and place to use antibiotics. However, he continued, "50 to 70 percent of antibiotic use in humans in North America does not fit that bill. It's not necessary. You can dial it back." You may need to be assertive in the conversation with your doctor regarding a course of antibiotics. If you want your microbiome to take care of you, you have to stand up for it!

The best way to be your own health advocate is to stay informed. Dr. Patrick gave the analogy of the rules for a good journalist: "Have your trusted sources. I often point patients towards WebMD and PubMed. The amount of unfiltered health information out there on the Internet is extremely troubling." In the evolving era of unreliable news, it's increasingly important to distinguish between credible sources of information and misinformed, biased—and even downright wrong—sources. If you are unsure about the accuracy of a medical claim, fact-check it against other studies and consult a health professional who may have more data than you can access.

> ***Recommendation:*** *When it comes to anything about your health, not just the microbiome, be proactive and start the conversation with your doctor so you can work together to keep you well for the duration of your (long) life.*

The Future Is Bright

The past decade has seen a remarkable renaissance in our understanding of the microbes inside and around us, from recognizing that we co-evolved with them to discovering the fundamental roles they play in both lifelong health and disease. The exploding knowledge opens many exciting new possibilities for how we can use microbes' many beneficial capacities in the future for human health.

The field is rapidly moving from "microbe cataloguing" (identifying exactly which microbes are in and around the body) to figuring out the microbial genes and mechanisms that are responsible for particular effects. As we define particular microbial species and their molecules, it opens up opportunities for novel treatments. For example, we are still in the early days of defining which components of diet particular microbes utilize. While we now recognize that fibre is converted into SCFAs with beneficial effects, this is just the beginning of what could be a regular partnership between nutrition and microbiology. Specific diet components will be linked to specific microbes, and even metabolic pathways and molecules within them, thus allowing diets to be tailored to individuals for specific defined health benefits. Recommended food guidelines may be adjusted to incorporate the microbiome (e.g., perhaps we will distinguish fermented foods as a new food group), and they are even more likely to be personalized for unique microbiomes and needs. This will include designing specific diets for each stage of life, particularly early in life when one's microbiome is being established and later in life to enhance health and longevity.

Genetic engineering has changed the entire biotechnology sector and will be applied to the microbiome field. We are already starting to see common probiotics engineered to enhance their benefits. However, microbes could be engineered to better regulate energy balance, to detoxify harmful toxic products, or synthesize beneficial molecules including vitamins. A recent major advance in molecular biology involved the development of CRISPR-Cas9, a tool that has made the replacement of genes in practically any organism much, much easier. This and other tools will make the engineering of beneficial microbes much more practical. We also sorely need tools to surgically alter microbiome composition, such that it can be shaped into a more beneficial, less dysbiotic community in individuals (especially older adults) when standard treatments for diseases are unsuccessful or impossible. We are already seeing companies performing early clinical trials that replenish dysbiotic microbiomes with a mixture of specific microbes grown in the laboratory. This opens many exciting avenues that could be used to repair poorly functioning microbiomes that don't respond to other methods.

With each generation, the collective microbiome of humans decreases in diversity and becomes more homogeneous as we share cities, food sources, and each other's company. As Dr. Martin Blaser points out in his book *Missing Microbes*, we are at significant risk of losing microbes that are part of our evolution, and, from what we know, this could have significant effects on our health (witness the rise of asthma, obesity, and the many diseases in our society correlated with a decrease in microbial diversity). The real problem is what to do about this. There are some groups that are working on biobanking microbes as a sort of "microzoo." As discussed previously, some current members of our microbiome really may be endangered species, which could have significant effects on humans. We are removing a key component to our evolution as a species, as we evolved with our microbiome.

Similar things can be contemplated about our microbes as we age. We saw in Chapter 2 the beneficial effect of skin microbes from younger women on skin health in older women. We know that our microbes shift as we get older, and, at least in mice, transplanting microbes from younger animals into older ones can decrease inflammation. One can imagine a future where you might biobank microbes from your youthful skin in a tube in a freezer, which would become part of your retirement package to boost skin appearance and overall health later in life. This personal biobank could also be useful if you were diagnosed with cancer and needed a bone marrow transplant, or if you suddenly got a *C. difficile* infection and needed to boost your microbiome at a critical time. We might also see the application of younger persons' microbes into commercially available products to cultivate the healthful effects of vigorous youthful microbes later in life.

In Chapter 8, we saw an example of "drugging the bugs" with respect to inhibiting TMA production by the gut microbiota (produced from choline and carnitine in animal products such as red meat and egg yolks) to decrease cardiovascular disease. In the future, as we establish the exact roles of microbes and their enzymes, we may have at our disposal a vast new set of pharmaceuticals. For example, *Eggerthella lenta* is a gut microbe that can block digoxin activity (a heart medication) by breaking down the drug. Researchers have identified the microbial pathway responsible and have now shown that they can inhibit this pathway, which could provide a new tool to enhance digoxin's effects in more people. One advantage of drugging the microbes, but not ourselves, is that specific adverse reactions to the drug shouldn't affect us. The drugs can act directly upon the microbes with limited side effects to our *H. sapiens* gene products.

This brings us to the effect of the microbiome on personalized medicine. Personalized medicine is based on the premise that different people respond differently to drugs, and that by understanding a person's genetic composition, one should be able to better predict which drugs should work with each individual. The problem with that theory is that we are 99.9 percent identical genetically (sex chromosomes excepted). Scientists have been able to identify genetic loci that dictate a particular drug's effect for only a small number of drugs. However, each of us has a unique microbiome, with only a fraction of shared microbes between people—even among those living together. It is the microbiome that gives individuals much of their uniqueness. When you swallow a drug, it is usually the microbes in the gut that break it down or modify it into a molecule that may cause adverse drug effects. In the future, microbiome sequencing may be even more helpful to doctors than genetic sequencing, as it could provide guidance about which drugs to use and at what dosage by considering your particular microbiome's composition and potential interactions.

For every bacterium it is estimated that there are about a hundred bacterial viruses called bacteriophages or phages. Studying the "virome" in the human gut is still new (but advancing rapidly, with lots of promising new findings). There are an estimated 1,200 viral genotypes in the gut. Each phage has a specific bacterial host, which prompts the idea that phages can be used to specifically target a bacterium. This forms the basis of "phage therapy," a technology pioneered in the Soviet Union in the 1950s and 1960s. However, this technology has problems, as bacteria rapidly become resistant to phages, and it has not withstood Western clinical trials. New uses for phages are on the horizon. For example, CRISP-Cas9 can be packaged into a phage and used to target a specific pathogen, modulate the microbiome, or even engineer a bacterium to treat a disease.

Looking further into the future, it may be possible to engineer microbes that increase longevity. *Caenorhabditis elegans* (frequently called *C. elegans*) is a small nematode (worm) that has been studied extensively in many labs. It feeds off of bacteria such as *E. coli* that also colonize its gut. When researchers made individual mutations in every non-essential *E. coli* gene (all 3,983 of them) and then fed them to worms, they identified twenty-nine bacterial genes that increased worm longevity. Some of the gene products affected worm pathways already known to increase longevity, including mitochondria and unfolded protein responses. Although very far from the world of humans, this work hints that there may be ways to alter the microbiome to enhance longevity. However, these types of experiments in humans would take many generations to reproduce given that the average lifespan of *C. elegans* is twelve to eighteen days, and average time between generations is four days!

Knowing the role of the microbiome in human health and disease has opened up many fascinating possibilities to harness this information to develop beneficial products. However, it is still hindered by the lack of mechanistic knowledge about exactly which microbes, which metabolic pathways, and which molecules are responsible for any particular activity. As science relentlessly progresses in these areas, countless opportunities exist to enter a whole new era of pharmacology based on microbes and their products.

In Pursuit of Health and Longevity: The Whole-Body Microbiome

So, what can you do now? First and foremost, recognize that we live in harmony with most of our microbes. They are a normal, essential part of everyday human life. Choose your lifestyle habits based not only *your* needs, but also your microbes'. Protocols of hygiene and health exist on a spectrum—from hand sanitizer and antibiotics to how you move and what you eat—and can all be calibrated to allow your microbes to thrive along with you. Recognize that as you age, you may need to adjust things to maintain a healthy diversity of microbes.

In the popular book *The Blue Zones*, Dan Buettner examined five regions of the world where people live much longer than average: Sardinia, Okinawa (Japan), Loma Linda (California), Costa Rica, and Greece. He observed and compared many aspects of their lives, including social habits, diet, and exercise. In Buettner's nine lessons learned from those with extraordinary longevity, we found that many are underscored by the microbiome. First, one's diet has obvious connections. These long-lived populations frequently ate plant-based foods, fish, nuts, legumes, and little red meat. The Japanese diet rich in fish and rice and the Mediterranean diets were expected, but southern California? In Loma Linda, there are many Seventh-Day Adventists, who are encouraged to follow a well-balanced vegetarian diet. Given all we have seen with diet and microbe interconnections, this certainly makes sense.

The second theme we noticed is exercise. One does not have to be a marathon runner, but countless studies find that staying active promotes health and longevity. Many in the Blue Zones were sheep herders or walked to the village daily for groceries and social contact. In southern California, the weather is conducive to outdoor activities such as jogging, tennis, and walking. Perhaps their longevity is impacted by active lifestyles that encourage beneficial microbes.

The third theme is strong social bonds with family and friends. This might include an eighty-year-old daughter caring for her one hundred-year-old father, or simply playing a daily game of cards. People in most of these areas of the world live among their extended families and in multigenerational households. The religious community in Loma Linda acts like an extended family with regular social contact. And we know that social contact is a great way to share and swap microbes. Unfortunately, modern Western societies tend to do the complete opposite in caring for older people. We put them in assisted living and nursing homes, feed them a limited diet, and only visit when we can find a spare moment in our busy lives. As you and loved ones age, keep these lessons in mind to maintain a vibrant and healthy lifestyle.

While science is still in the shallow end of this vast pool of exploration, the whole-body microbiome represents a promising new frontier of aging research—a fountain of youth that's far bigger than we ever could have imagined. Keep an eye out for exciting new scientific advances in "microbe therapies" that we saw glimpses of in this book. At thirty and sixty years old, we (Jessica and Brett) regularly consider and want to proactively address our own age-related concerns—from wrinkles and declining mobility to Alzheimer's disease. We have adopted many of the diet and lifestyle strategies suggested throughout this book, pursuing these strategies with the knowledge that the more we can learn about their engagement with the microbiome, the more we can improve quality of life in our globally aging population. As new discoveries appear every day, we may soon realize that the fountain of youth we've travelled far and wide looking for is right under (and inside, and on) our own noses. We hope that you and your microbes may live long—and prosper.

Acknowledgements

This book is testament to the enthusiasm and generosity of a large number of people. We were extremely fortunate to receive input from many talented individuals who shared their perspectives, experiences, and collective knowledge.

We thank all of the scientific and medical experts who provided interviews and edits for each chapter: Marty Blaser, Shoki Dedhar, Eran Elinav, Richard Ellen, Stan Hazen, Greg Hillebrand, Jim Hogg, Dan Littman, Brian MacVicar, Anne Martin-Matthews, Heather McKay, Dave Patrick, Mary Ellen Sanders, and Lynn Stothers. They helped ensure that we weren't misinterpreting the scientific data in all our enthusiasm for this burgeoning area. Their wide array of knowledge—most are not experts in microbiology—was fundamental to the framing and content of chapters, and gave us insight into the many ways that microbiology can become more broadly incorporated throughout healthcare and daily pursuits of well-being.

Numerous colleagues, friends, and family members read excerpts, provided comments, and sent us relevant articles along the way. This includes Hazel and Wallace Allen, Claire Arrieta, Kylynda Bauer, Monica Bennington, John Bienenstock, Mihai Cirstea, Silke Cresswell, Liam Finlay, Derek Gregory, Natasa Jovic, Malcolm Kendall, Roderick MacDonald, Patti Martin, Shaylih Muehlmann, Janet Rossant, and Hillary

Waters. We particularly thank Janis Sarra for reading an early draft of the book and providing so many valuable comments.

We thank our outstanding agents, Chris Casuccio and John Pearce. Our editor, Jennifer Kurdyla, provided wonderful comments, edits, and compelling questions throughout the writing process. We appreciate her crucial role in helping us make this book accessible and useful to people of all walks and ages.

Finally, we thank our spouses, Jane and Matt, who were involved in endless discussions, helped brainstorm and "gut-check" ideas, read drafts, and provided countless articles and additional information. We are forever grateful for their unfailing encouragement, patience, and love.

Selected References

1. THE FOUNTAIN OF YOUTH IS FULL OF . . . MICROBES?

Biagi, E., Candela, M., Franceschi, C., & Brigidi, P. (2011). The aging gut microbiota: new perspectives. *Ageing research reviews, 10*(4), 428–29. doi:10.1016/j.arr.2011.03.004

Biagi, E., Nylund, L., Candela, M., Ostan, R., Bucci, L., Pini, E., . . . De Vos, W. (2010). Through ageing, and beyond: gut microbiota and inflammatory status in seniors and centenarians. *PloS one, 5*(5), e10667. doi:10.1371/journal.pone.0010667

Brüssow, H. (2013). Microbiota and healthy ageing: observational and nutritional intervention studies. *Microbial biotechnology, 6*(4), 326–34. doi:10.1111/1751-7915.12048

Claesson, M. J., Jeffery, I. B., Conde, S., Power, S. E., O'Connor, E. M., Cusack, S., . . . O'Toole, P. W. (2012). Gut microbiota composition correlates with diet and health in the elderly. *Nature, 488*(7410), 178–84. doi:10.1038/nature11319

Gilbert, J. A., Blaser, M. J., Caporaso, J. G., Jansson, J. K., Lynch, S. V., & Knight, R. (2018). Current understanding of the human microbiome. *Nature medicine, 24*(4), 392–400. doi:10.1038/nm.4517

Heintz, C., & Mair, W. (2014). You are what you host: microbiome modulation of the aging process. *Cell, 156*(3), 408–11. doi:10.1016/j.cell.2014.01.025

Jackson, M. A., Jeffery, I. B., Beaumont, M., Bell, J. T., Clark, A. G., Ley, R. E., . . . Steves, C. J. (2016). Signatures of early frailty in the gut microbiota. *Genome Medicine, 8*(1), 8. doi:10.1186/s13073-016-0262-7

Kong, F., Hua, Y., Zeng, B., Ning, R., Li, Y., & Zhao, J. (2016). Gut microbiota signatures of longevity. *Current Biology: CB, 26*(18), R832–R833. doi:10.1016/j.cub.2016.08.015

Lynch, D. B., Jeffery, I. B., & O'Toole, P. W. (2015). The role of the microbiota in ageing: current state and perspectives. *Wiley interdisciplinary reviews. Systems Biology and Medicine, 7*(3), 131–38. doi:10.1002/wsbm.1293

Rondanelli, M., Giacosa, A., Faliva, M. A., Perna, S., Allieri, F., & Castellazzi, A. M. (2015). Review on microbiota and effectiveness of probiotics use in older. *World Journal of Clinical Cases, 3*(2), 156–62. doi:10.12998/wjcc.v3.i2.156

Saraswati, S., & Sitaraman, R. (2014). Aging and the human gut microbiota-from correlation to causality. *Frontiers in Microbiology, 5*(764). doi:10.3389/fmicb.2014.00764

Smits, S. A., Leach, J., Sonnenburg, E. D., Gonzalez, C. G., Lichtman, J. S., Reid, G., ... Sonnenburg, J. L. (2017). Seasonal cycling in the gut microbiome of the Hadza hunter-gatherers of Tanzania. *Science, 357*(6353), 802–6. doi:10.1126/science.aan4834

Thevaranjan, N., Puchta, A., Schulz, C., Naidoo, A., Szamosi, J. C., Verschoor, C. P., ... Bowdish, D. M. E. (2017). Age-Associated Microbial Dysbiosis Promotes Intestinal Permeability, Systemic Inflammation, and Macrophage Dysfunction. *Cell Host & Microbe, 21*(4), 455–66 e454. doi:10.1016/j.chom.2017.03.002

Tiihonen, K., Ouwehand, A. C., & Rautonen, N. (2010). Human intestinal microbiota and healthy ageing. *Ageing Research Reviews, 9*(2), 107–16. doi:10.1016/j.arr.2009.10.004

Zapata, H. J., & Quagliarello, V. J. (2015). The microbiota and microbiome in aging: potential implications in health and age-related diseases. *Journal of the American Geriatrics Society, 63*(4), 776–81. doi:10.1111/jgs.13310

2. YOUR MICROBES ARE GLOWING: THE SKIN MICROBIOME

Grice, E. A., & Segre, J. A. (2011). The skin microbiome. Nature reviews. *Microbiology, 9*(4), 244–53. doi:10.1038/nrmicro2537

Lax, S., Hampton-Marcell, J. T., Gibbons, S. M., Colares, G. B., Smith, D., Eisen, J. A., & Gilbert, J. A. (2015). Forensic analysis of the microbiome of phones and shoes. *Microbiome, 3*(21). doi:10.1186/s40168-015-0082-9

Lee, D. E., Huh, C. S., Ra, J., Choi, I. D., Jeong, J. W., Kim, S. H., ... Ahn, Y. T. (2015). Clinical Evidence of Effects of Lactobacillus plantarum HY7714 on Skin Aging: A Randomized, Double Blind, Placebo-Controlled Study. *Journal of Microbiology and Biotechnology, 25*(12), 2160–68. doi:10.4014/jmb.1509.09021

Levkovich, T., Poutahidis, T., Smillie, C., Varian, B. J., Ibrahim, Y. M., Lakritz, J. R., ... Erdman, S. E. (2013). Probiotic bacteria induce a 'glow of health.' *PloS One, 8*(1), e53867. doi:10.1371/journal.pone.0053867

Nodake, Y., Matsumoto, S., Miura, R., Honda, H., Ishibashi, G., Dekio, I., & Sakakibara, R. (2015). Pilot study on novel skin care method by augmentation with Staphylococcus epidermidis, an autologous skin microbe—A blinded randomized clinical trial. *Journal of Dermatological Science, 79*(2), 119–26. doi:10.1016/j.jdermsci.2015.05.001

Oh, J., Byrd, A. L., Park, M., Kong, H. H., & Segre, J. A. (2016). Temporal Stability of the Human Skin Microbiome. *Cell, 165*(4), 854–66. doi:10.1016/j.cell.2016.04.008

Shin, H., Price, K., Albert, L., Dodick, J., Park, L., & Dominguez-Bello, M. G. (2016). Changes in the Eye Microbiota Associated with Contact Lens Wearing. *mBio, 7*(2), e00198. doi:10.1128/mBio.00198-16

3. MIND YOUR MICROBES: MICROBES AND THE BRAIN

Foster, J. A., & McVey Neufeld, K. A. (2013). Gut-brain axis: how the microbiome influences anxiety and depression. *Trends in Neurosciences, 36*(5), 305–12. doi:10.1016/j.tins.2013.01.005

Hsiao, E. Y., McBride, S. W., Hsien, S., Sharon, G., Hyde, E. R., McCue, T., . . . Mazmanian, S. K. (2013). Microbiota modulate behavioral and physiological abnormalities associated with neurodevelopmental disorders. *Cell, 155*(7), 1451–63. doi:10.1016/j.cell.2013.11.024

Leone, V., Gibbons, S. M., Martinez, K., Hutchison, A. L., Huang, E. Y., Cham, C. M., . . . Chang, E. B. (2015). Effects of diurnal variation of gut microbes and high-fat feeding on host circadian clock function and metabolism. *Cell Host & Microbe, 17*(5), 681–89. doi:10.1016/j.chom.2015.03.006

Lurie, I., Yang, Y. X., Haynes, K., Mamtani, R., & Boursi, B. (2015). Antibiotic exposure and the risk for depression, anxiety, or psychosis: a nested case-control study. *The Journal of Clinical Psychiatry, 76*(11), 1522–28. doi:10.4088/JCP.15m09961

Messaoudi, M., Lalonde, R., Violle, N., Javelot, H., Desor, D., Nejdi, A., . . . Cazaubiel, J. M. (2011). Assessment of psychotropic-like properties of a probiotic formulation (Lactobacillus helveticus R0052 and Bifidobacterium longum R0175) in rats and human subjects. *The British Journal of Nutrition, 105*(5), 755–64. doi:10.1017/S0007114510004319

Morris, M. C., Tangney, C. C., Wang, Y., Sacks, F. M., Bennett, D. A., & Aggarwal, N. T. (2015). MIND diet associated with reduced incidence of Alzheimer's disease. *Alzheimer's & Dementia: the Journal of the Alzheimer's Association, 11*(9), 1007–14. doi:10.1016/j.jalz.2014.11.009

Sampson, T. R., Debelius, J. W., Thron, T., Janssen, S., Shastri, G. G., Ilhan, Z. E., . . . Mazmanian, S. K. (2016). Gut Microbiota Regulate Motor Deficits and Neuroinflammation in a Model of Parkinson's Disease. *Cell, 167*(6), 1469–80 e1412. doi:10.1016/j.cell.2016.11.018

Sampson, T. R., & Mazmanian, S. K. (2015). Control of brain development, function, and behavior by the microbiome. *Cell Host & Microbe, 17*(5), 565–76. doi:10.1016/j.chom.2015.04.011

Svensson, E., Horvath-Puho, E., Thomsen, R. W., Djurhuus, J. C., Pedersen, L., Borghammer, P., & Sorensen, H. T. (2015). Vagotomy and subsequent risk of Parkinson's disease. *Annals of Neurology, 78*(4), 522–29. doi:10.1002/ana.24448

Thaiss, C. A., Zeevi, D., Levy, M., Zilberman-Schapira, G., Suez, J., Tengeler, A. C., . . . Elinav, E. (2014). Transkingdom control of microbiota diurnal oscillations promotes metabolic homeostasis. *Cell, 159*(3), 514–29. doi:10.1016/j.cell.2014.09.048

Vogt, N. M., Kerby, R. L., Dill-McFarland, K. A., Harding, S. J., Merluzzi, A. P., Johnson, S. C., . . . Rey, F. E. (2017). Gut microbiome alterations in Alzheimer's disease. *Scientific Reports, 7*(1), 13537. doi:10.1038/s41598-017-13601-y

4. HEALTHY SMILE, HEALTHY YOU: THE ORAL MICROBIOME

Curtis, M. A., Zenobia, C., & Darveau, R. P. (2011). The relationship of the oral microbiota to periodontal health and disease. *Cell Host & Microbe, 10*(4), 302–6. doi:10.1016/j.chom.2011.09.008

Shoemark, D. K., & Allen, S. J. (2015). The microbiome and disease: reviewing the links between the oral microbiome, aging, and Alzheimer's disease. *Journal of Alzheimer's Disease : JAD, 43*(3), 725–38. doi:10.3233/JAD-141170

Wade, W. G. (2013). The oral microbiome in health and disease. *Pharmacological Research, 69*(1), 137–43. doi:10.1016/j.phrs.2012.11.006

Weyrich, L. S., Dobney, K., & Cooper, A. (2015). Ancient DNA analysis of dental calculus. *Journal of Human Evolution, 79*, 119–24. doi:10.1016/j.jhevol.2014.06.018

Zarco, M. F., Vess, T. J., & Ginsburg, G. S. (2012). The oral microbiome in health and disease and the potential impact on personalized dental medicine. *Oral Diseases, 18*(2), 109–20. doi:10.1111/j.1601-0825.2011.01851

5. TAKE A DEEP BREATH: THE LUNG MICROBIOME

Dickson, R. P., Martinez, F. J., & Huffnagle, G. B. (2014). The role of the microbiome in exacerbations of chronic lung diseases. *Lancet, 384*(9944), 691–702. doi:10.1016/S0140-6736(14)61136-3

Faner, R., Sibila, O., Agusti, A., Bernasconi, E., Chalmers, J. D., Huffnagle, G. B., ... Monso, E. (2017). The microbiome in respiratory medicine: current challenges and future perspectives. *The European Respiratory Journal, 49*(4). doi:10.1183/13993003.02086-2016

Segal, L. N., & Blaser, M. J. (2014). A brave new world: the lung microbiota in an era of change. *Annals of the American Thoracic Society, 11* Suppl 1, S21-27. doi:10.1513/AnnalsATS.201306-189MG

Sze, M. A., Dimitriu, P. A., Suzuki, M., McDonough, J. E., Campbell, J. D., Brothers, J. F., ... Hogg, J. C. (2015). Host Response to the Lung Microbiome in Chronic Obstructive Pulmonary Disease. *American Journal of Respiratory and Critical Care Medicine, 192*(4), 438–45. doi:10.1164/rccm.201502-0223OC

6. BELLY BUGS: THE STOMACH MICROBIOME

Homan, M., & Orel, R. (2015). Are probiotics useful in Helicobacter pylori eradication? *World Journal of Gastroenterology, 21*(37), 10644–53. doi:10.3748/wjg.v21.i37.10644

Jackson, M. A., Goodrich, J. K., Maxan, M. E., Freedberg, D. E., Abrams, J. A., Poole, A. C., ... Steves, C. J. (2016). Proton pump inhibitors alter the composition of the gut microbiota. *Gut, 65*(5), 749–56. doi:10.1136/gutjnl-2015-310861

Kienesberger, S., Cox, L. M., Livanos, A., Zhang, X. S., Chung, J., Perez-Perez, G. I., ... Blaser, M. J. (2016). Gastric Helicobacter pylori Infection Affects Local and

Distant Microbial Populations and Host Responses. *Cell Reports, 14*(6), 1395–1407. doi:10.1016/j.celrep.2016.01.017

Yang, I., Nell, S., & Suerbaum, S. (2013). Survival in hostile territory: the microbiota of the stomach. *FEMS Microbiology Reviews, 37*(5), 736–61. doi:10.1111/1574-6976.12027

Yang, I., Woltemate, S., Piazuelo, M. B., Bravo, L. E., Yepez, M. C., Romero-Gallo, J., . . . Suerbaum, S. (2016). Different gastric microbiota compositions in two human populations with high and low gastric cancer risk in Colombia. *Scientific Reports, 6*(18594). doi:10.1038/srep18594

7. MICROBE MECCA: THE GUT MICROBIOME

Bala, S., Marcos, M., Gattu, A., Catalano, D., & Szabo, G. (2014). Acute binge drinking increases serum endotoxin and bacterial DNA levels in healthy individuals. *PloS One, 9*(5), e96864. doi:10.1371/journal.pone.0096864

Baothman, O. A., Zamzami, M. A., Taher, I., Abubaker, J., & Abu-Farha, M. (2016). The role of Gut Microbiota in the development of obesity and Diabetes. *Lipids in Health and Disease, 15*(108). doi:10.1186/s12944-016-0278-4

Biagi, E., Franceschi, C., Rampelli, S., Severgnini, M., Ostan, R., Turroni, S., . . . Candela, M. (2016). Gut Microbiota and Extreme Longevity. *Current Biology: CB, 26*(11), 1480–85. doi:10.1016/j.cub.2016.04.016

Bian, G., Gloor, G. B., Gong, A., Jia, C., Zhang, W., Hu, J., . . . Li, J. (2017). The Gut Microbiota of Healthy Aged Chinese Is Similar to That of the Healthy Young. *mSphere, 2*(5). doi:10.1128/mSphere.00327-17

Brandt, L. J. (2017). Fecal Microbiota Therapy With a Focus on Clostridium difficile Infection. *Psychosomatic Medicine, 79*(8), 868–73. doi:10.1097/PSY.0000000000000511

Brüssow, H. (2013). Microbiota and healthy ageing: observational and nutritional intervention studies. *Microbial Biotechnology, 6*(4), 326–34. doi:10.1111/1751-7915.12048

Chu, H., Khosravi, A., Kusumawardhani, I. P., Kwon, A. H., Vasconcelos, A. C., Cunha, L. D., . . . Mazmanian, S. K. (2016). Gene-microbiota interactions contribute to the pathogenesis of inflammatory bowel disease. *Science, 352*(6289), 1116–20. doi:10.1126/science.aad9948

Deehan, E. C., & Walter, J. (2016). The Fiber Gap and the Disappearing Gut Microbiome: Implications for Human Nutrition. *Trends in Endocrinology and Metabolism: TEM, 27*(5), 239–42. doi:10.1016/j.tem.2016.03.001

Forslund, K., Hildebrand, F., Nielsen, T., Falony, G., Le Chatelier, E., Sunagawa, S., . . . Pedersen, O. (2015). Disentangling type 2 diabetes and metformin treatment signatures in the human gut microbiota. *Nature, 528*(7581), 262–66. doi:10.1038/nature15766

Iyer, N., & Vaishnava, S. (2016). Alcohol Lowers Your (Intestinal) Inhibitions. *Cell Host & Microbe, 19*(2), 131–33. doi:10.1016/j.chom.2016.01.014

Jayasinghe, T. N., Chiavaroli, V., Holland, D. J., Cutfield, W. S., & O'Sullivan, J. M. (2016). The New Era of Treatment for Obesity and Metabolic Disorders: Evidence and Expectations for Gut Microbiome Transplantation. *Frontiers in cellular and Infection Microbiology, 6*(15). doi:10.3389/fcimb.2016.00015

Kong, F., Hua, Y., Zeng, B., Ning, R., Li, Y., & Zhao, J. (2016). Gut microbiota signatures of longevity. *Current Biology: CB, 26*(18), R832–R833. doi:10.1016/j.cub.2016.08.015

Konig, J., Siebenhaar, A., Hogenauer, C., Arkkila, P., Nieuwdorp, M., Noren, T., . . . Brummer, R. J. (2017). Consensus report: faecal microbiota transfer - clinical applications and procedures. *Alimentary Pharmacology & Therapeutics, 45*(2), 222–39. doi:10.1111/apt.13868

Leung, C., Rivera, L., Furness, J. B., & Angus, P. W. (2016). The role of the gut microbiota in NAFLD. *Nature reviews. Gastroenterology & Hepatology, 13*(7), 412–25. doi:10.1038/nrgastro.2016.85

Maher, R. L., Hanlon, J., & Hajjar, E. R. (2014). Clinical consequences of polypharmacy in elderly. *Expert Opinion on Drug Safety, 13*(1), 57–65. doi:10.1517/14740338.2013.827660

Santoro, A., Ostan, R., Candela, M., Biagi, E., Brigidi, P., Capri, M., & Franceschi, C. (2018). Gut microbiota changes in the extreme decades of human life: a focus on centenarians. *Cellular and Molecular Life Sciences: CMLS, 75*(1), 129–48. doi:10.1007/s00018-017-2674-y

Sonnenburg, E. D., Smits, S. A., Tikhonov, M., Higginbottom, S. K., Wingreen, N. S., & Sonnenburg, J. L. (2016). Diet-induced extinctions in the gut microbiota compound over generations. *Nature, 529*(7585), 212–15. doi:10.1038/nature16504

Suez, J., Korem, T., Zeevi, D., Zilberman-Schapira, G., Thaiss, C. A., Maza, O., . . . Elinav, E. (2014). Artificial sweeteners induce glucose intolerance by altering the gut microbiota. *Nature, 514*(7521), 181–86. doi:10.1038/nature13793

Thaiss, C. A., Itav, S., Rothschild, D., Meijer, M., Levy, M., Moresi, C., . . . Elinav, E. (2016). Persistent microbiome alterations modulate the rate of post-dieting weight regain. *Nature, 540*, 544–51. doi:10.1038/nature20796

Zeevi, D., Korem, T., Zmora, N., Israeli, D., Rothschild, D., Weinberger, A., . . . Segal, E. (2015). Personalized Nutrition by Prediction of Glycemic Responses. *Cell, 163*(5), 1079–94. doi:10.1016/j.cell.2015.11.001

8. LOVE BUGS: THE HEART AND THE MICROBIOME

American Heart Association. (2017). Good Fats and Bad Fats: The Facts on Healthy Fats. Retrieved from healthyforgood.heart.org/eat-smart/infographics/the-facts-on-fats

Bala, S., Marcos, M., Gattu, A., Catalano, D., & Szabo, G. (2014). Acute binge drinking increases serum endotoxin and bacterial DNA levels in healthy individuals. *PloS One, 9*(5), e96864. doi:10.1371/journal.pone.0096864

Chen, M. L., Yi, L., Zhang, Y., Zhou, X., Ran, L., Yang, J., . . . Mi, M. T. (2016). Resveratrol Attenuates Trimethylamine-N-Oxide (TMAO)-Induced Atherosclerosis by

Regulating TMAO Synthesis and Bile Acid Metabolism via Remodeling of the Gut Microbiota. *mBio, 7*(2), e02210-02215. doi:10.1128/mBio.02210-15

Clarke, S. F., Murphy, E. F., O'Sullivan, O., Lucey, A. J., Humphreys, M., Hogan, A., ... Cotter, P. D. (2014). Exercise and associated dietary extremes impact on gut microbial diversity. *Gut, 63*(12), 1913–20. doi:10.1136/gutjnl-2013-306541

Gregory, J. C., Buffa, J. A., Org, E., Wang, Z., Levison, B. S., Zhu, W., ... Hazen, S. L. (2015). Transmission of atherosclerosis susceptibility with gut microbial transplantation. *The Journal of Biological Chemistry, 290*(9), 5647–60. doi:10.1074/jbc.M114.618249

Koeth, R. A., Wang, Z., Levison, B. S., Buffa, J. A., Org, E., Sheehy, B. T., ... Hazen, S. L. (2013). Intestinal microbiota metabolism of L-carnitine, a nutrient in red meat, promotes atherosclerosis. *Nature Medicine, 19*(5), 576–85. doi:10.1038/nm.3145

Teicholz, N. (2014). *The Big Fat Surprise: Why Butter, Meat and Cheese Belong in a Healthy Diet.* New York, NY: Simon & Schuster Paperbacks.

Wang, Z., Roberts, A. B., Buffa, J. A., Levison, B. S., Zhu, W., Org, E., ... Hazen, S. L. (2015). Non-lethal Inhibition of Gut Microbial Trimethylamine Production for the Treatment of Atherosclerosis. *Cell, 163*(7), 1585–95. doi:10.1016/j.cell.2015.11.055

Zhu, W., Gregory, J. C., Org, E., Buffa, J. A., Gupta, N., Wang, Z., ... Hazen, S. L. (2016). Gut Microbial Metabolite TMAO Enhances Platelet Hyperreactivity and Thrombosis Risk. *Cell, 165*(1), 111–24. doi:10.1016/j.cell.2016.02.011

9. FEMALES ARE NOT SMALL MALES: MENOPAUSE AND THE VAGINAL MICROBIOME

Baker, J. M., Al-Nakkash, L., & Herbst-Kralovetz, M. M. (2017). Estrogen-gut microbiome axis: Physiological and clinical implications. *Maturitas, 103*, 45–53. doi:10.1016/j.maturitas.2017.06.025

Ma, B., Forney, L. J., & Ravel, J. (2012). Vaginal microbiome: rethinking health and disease. *Annual Review of Microbiology, 66*, 371–89. doi:10.1146/annurev-micro-092611-150157

Martin, D. H. (2012). The microbiota of the vagina and its influence on women's health and disease. *The American Journal of the Medical Sciences, 343*(1), 2–9. doi:10.1097/MAJ.0b013e31823ea228

Muhleisen, A. L., & Herbst-Kralovetz, M. M. (2016). Menopause and the vaginal microbiome. *Maturitas, 91*, 42–50. doi:10.1016/j.maturitas.2016.05.015

Schroder, W., Sommer, H., Gladstone, B. P., Foschi, F., Hellman, J., Evengard, B., & Tacconelli, E. (2016). Gender differences in antibiotic prescribing in the community: a systematic review and meta-analysis. *The Journal of Antimicrobial Chemotherapy, 71*(7), 1800–1806. doi:10.1093/jac/dkw054

Schwenger, E. M., Tejani, A. M., & Loewen, P. S. (2015). Probiotics for preventing urinary tract infections in adults and children. *The Cochrane Database of Systematic Reviews* (12), CD008772. doi:10.1002/14651858.CD008772.pub2

Shen, J., Song, N., Williams, C. J., Brown, C. J., Yan, Z., Xu, C., & Forney, L. J. (2016). Effects of low dose estrogen therapy on the vaginal microbiomes of women with atrophic vaginitis. *Scientific Reports, 6*(24380). doi:10.1038/srep24380

Thomas-White, K., Brady, M., Wolfe, A. J., & Mueller, E. R. (2016). The bladder is not sterile: History and current discoveries on the urinary microbiome. *Current Bladder Dysfunction Reports, 11*(1), 18–24. doi:10.1007/s11884-016-0345-8

Younes, J. A., Lievens, E., Hummelen, R., van der Westen, R., Reid, G., & Petrova, M. I. (2018). Women and Their Microbes: The Unexpected Friendship. *Trends in Microbiology*, 26(1), 16–32. doi:10.1016/j.tim.2017.07.008

10. MICROBES MEET CANCER

Bacchus, C. M., Dunfield, L., Gorber, S. C., Holmes, N. M., Birtwhistle, R., Dickinson, J. A., . . . Tonelli, M. (2016). Recommendations on screening for colorectal cancer in primary care. *CMAJ : Canadian Medical Association journal = journal de l'Association medicale canadienne, 188*(5), 340–48. doi:10.1503/cmaj.151125

Cao, Y., Wu, K., Mehta, R., Drew, D. A., Song, M., Lochhead, P., . . . Chan, A. T. (2018). Long-term use of antibiotics and risk of colorectal adenoma. *Gut, 67*(4), 672–78. doi:10.1136/gutjnl-2016-313413

Garrett, W. S. (2015). Cancer and the microbiota. *Science, 348*(6230), 80–86. doi:10.1126/science.aaa4972

Gopalakrishnan, V., Spencer, C. N., Nezi, L., Reuben, A., Andrews, M. C., Karpinets, T. V., . . . Wargo, J. A. (2018). Gut microbiome modulates response to anti-PD-1 immunotherapy in melanoma patients. *Science, 359*(6371), 97–103. doi:10.1126/science.aan4236

Hullar, M. A., Burnett-Hartman, A. N., & Lampe, J. W. (2014). Gut microbes, diet, and cancer. *Cancer Treatment and Research, 159*, 377–99. doi:10.1007/978-3-642-38007-5_22

Kostic, A. D., Chun, E., Robertson, L., Glickman, J. N., Gallini, C. A., Michaud, M., . . . Garrett, W. S. (2013). Fusobacterium nucleatum potentiates intestinal tumorigenesis and modulates the tumor-immune microenvironment. *Cell Host & Microbe, 14*(2), 207–15. doi:10.1016/j.chom.2013.07.007

Routy, B., Le Chatelier, E., Derosa, L., Duong, C. P. M., Alou, M. T., Daillere, R., . . . Zitvogel, L. (2018). Gut microbiome influences efficacy of PD-1-based immunotherapy against epithelial tumors. *Science, 359*(6371), 91–97. doi:10.1126/science.aan3706

Roy, S., & Trinchieri, G. (2017). Microbiota: a key orchestrator of cancer therapy. *Nature reviews. Cancer, 17*(5), 271–85. doi:10.1038/nrc.2017.13

Shono, Y., Docampo, M. D., Peled, J. U., Perobelli, S. M., Velardi, E., Tsai, J. J., . . . Jenq, R. R. (2016). Increased GVHD-related mortality with broad-spectrum antibiotic use after allogeneic hematopoietic stem cell transplantation in human patients and mice. *Science Translational Medicine, 8*(339), 339ra371. doi:10.1126/scitranslmed.aaf2311

Sivan, A., Corrales, L., Hubert, N., Williams, J. B., Aquino-Michaels, K., Earley, Z. M., . . . Gajewski, T. F. (2015). Commensal Bifidobacterium promotes antitumor immunity and facilitates anti-PD-L1 efficacy. *Science, 350*(6264), 1084–89. doi:10.1126/science.aac4255

Urbaniak, C., Gloor, G. B., Brackstone, M., Scott, L., Tangney, M., & Reid, G. (2016). The Microbiota of Breast Tissue and Its Association with Breast Cancer. *Applied and Environmental Microbiology, 82*(16), 5039–48. doi:10.1128/AEM.01235-16

Vetizou, M., Pitt, J. M., Daillere, R., Lepage, P., Waldschmitt, N., Flament, C., ... Zitvogel, L. (2015). Anticancer immunotherapy by CTLA-4 blockade relies on the gut microbiota. *Science, 350*(6264), 1079–84. doi:10.1126/science.aad1329

Weber, D., Jenq, R. R., Peled, J. U., Taur, Y., Hiergeist, A., Koestler, J., ... Holler, E. (2017). Microbiota Disruption Induced by Early Use of Broad-Spectrum Antibiotics Is an Independent Risk Factor of Outcome after Allogeneic Stem Cell Transplantation. *Biology of Blood and Marrow Transplantation: Journal of the American Society for Blood and Marrow Transplantation, 23*(5), 845-852. doi:10.1016/j.bbmt.2017.02.006

Yang, Y., Xia, Y., Chen, H., Hong, L., Feng, J., Yang, J., ... Ma, Y. (2016). The effect of perioperative probiotics treatment for colorectal cancer: short-term outcomes of a randomized controlled trial. *Oncotarget, 7*(7), 8432–40. doi:10.18632/oncotarget.7045

Zitvogel, L., Daillere, R., Roberti, M. P., Routy, B., & Kroemer, G. (2017). Anticancer effects of the microbiome and its products. *Nature reviews. Microbiology, 15*(8), 465–78. doi:10.1038/nrmicro.2017.44

11. MICROBE TUG-OF-WAR: THE IMMUNE SYSTEM

Atarashi, K., Tanoue, T., Oshima, K., Suda, W., Nagano, Y., Nishikawa, H., ... Honda, K. (2013). Treg induction by a rationally selected mixture of Clostridia strains from the human microbiota. *Nature, 500*(7461), 232–36. doi:10.1038/nature12331

Berer, K., Gerdes, L. A., Cekanaviciute, E., Jia, X., Xiao, L., Xia, Z., ... Wekerle, H. (2017). Gut microbiota from multiple sclerosis patients enables spontaneous autoimmune encephalomyelitis in mice. *Proceedings of the National Academy of Sciences of the United States of America, 114*(40), 10719–24. doi:10.1073/pnas.1711233114

Caballero, S., & Pamer, E. G. (2015). Microbiota-mediated inflammation and antimicrobial defense in the intestine. *Annual Review of Immunology, 33*, 227–56. doi:10.1146/annurev-immunol-032713-120238

Calder, P. C., Bosco, N., Bourdet-Sicard, R., Capuron, L., Delzenne, N., Dore, J., ... Visioli, F. (2017). Health relevance of the modification of low-grade inflammation in ageing (inflammageing) and the role of nutrition. *Ageing Research Reviews, 40*, 95–119. doi:10.1016/j.arr.2017.09.001

Cekanaviciute, E., Yoo, B. B., Runia, T. F., Debelius, J. W., Singh, S., Nelson, C. A., ... Baranzini, S. E. (2017). Gut bacteria from multiple sclerosis patients modulate human T cells and exacerbate symptoms in mouse models. *Proceedings of the National Academy of Sciences of the United States of America, 114*(40), 10713–18. doi:10.1073/pnas.1711235114

Cervantes-Barragan, L., Chai, J. N., Tianero, M. D., Di Luccia, B., Ahern, P. P., Merriman, J., . . . Colonna, M. (2017). Lactobacillus reuteri induces gut intraepithelial CD4(+)CD8alphaalpha(+) T cells. *Science, 357*(6353), 806–10. doi:10.1126/science.aah5825

Furusawa, Y., Obata, Y., Fukuda, S., Endo, T. A., Nakato, G., Takahashi, D., . . . Ohno, H. (2013). Commensal microbe-derived butyrate induces the differentiation of colonic regulatory T cells. *Nature, 504*(7480), 446–50. doi:10.1038/nature12721

Gill, N., & Finlay, B. B. (2011). The gut microbiota: challenging immunology. *Nature Reviews. Immunology, 11*(10), 636–37. doi:10.1038/nri3061

Honda, K., & Littman, D. R. (2016). The microbiota in adaptive immune homeostasis and disease. *Nature, 535*(7610), 75–84. doi:10.1038/nature18848

Hooper, L. V., Littman, D. R., & Macpherson, A. J. (2012). Interactions between the microbiota and the immune system. *Science, 336*(6086), 1268–73. doi:10.1126/science.1223490

Ivanov, I. I., Frutos Rde, L., Manel, N., Yoshinaga, K., Rifkin, D. B., Sartor, R. B., . . . Littman, D. R. (2008). Specific microbiota direct the differentiation of IL-17-producing T-helper cells in the mucosa of the small intestine. *Cell Host & Microbe, 4*(4), 337–49. doi:10.1016/j.chom.2008.09.009

Mazmanian, S. K., Liu, C. H., Tzianabos, A. O., & Kasper, D. L. (2005). An immunomodulatory molecule of symbiotic bacteria directs maturation of the host immune system. *Cell, 122*(1), 107–18. doi:10.1016/j.cell.2005.05.007

Postler, T. S., & Ghosh, S. (2017). Understanding the Holobiont: How Microbial Metabolites Affect Human Health and Shape the Immune System. *Cell Metabolism, 26*(1), 110–30. doi:10.1016/j.cmet.2017.05.008

Salazar, N., Arboleya, S., Valdes, L., Stanton, C., Ross, P., Ruiz, L., . . . de Los Reyes-Gavilan, C. G. (2014). The human intestinal microbiome at extreme ages of life. Dietary intervention as a way to counteract alterations. *Frontiers in Genetics, 5*(406). doi:10.3389/fgene.2014.00406

Scher, J. U., Sczesnak, A., Longman, R. S., Segata, N., Ubeda, C., Bielski, C., . . . Littman, D. R. (2013). Expansion of intestinal Prevotella copri correlates with enhanced susceptibility to arthritis. *eLife, 2*, e01202. doi:10.7554/eLife.01202

Thevaranjan, N., Puchta, A., Schulz, C., Naidoo, A., Szamosi, J. C., Verschoor, C. P., . . . Bowdish, D. M. E. (2017). Age-Associated Microbial Dysbiosis Promotes Intestinal Permeability, Systemic Inflammation, and Macrophage Dysfunction. *Cell Host & Microbe, 21*(4), 455–66 e454. doi:10.1016/j.chom.2017.03.002

Zhong, D., Wu, C., Zeng, X., & Wang, Q. (2018). The role of gut microbiota in the pathogenesis of rheumatic diseases. *Clinical Rheumatology, 37*(1), 25–34. doi:10.1007/s10067-017-3821-4

12. FLEX YOUR MICROBES: THE MUSCULOSKELETAL SYSTEM

Allen, J. M., Mailing, L. J., Niemiro, G. M., Moore, R., Cook, M. D., White, B. A., . . . Woods, J. A. (2018). Exercise Alters Gut Microbiota Composition and Function in Lean and Obese Humans. *Medicine and Science in Sports and Exercise, 50*(4), 747–57. doi:10.1249/MSS.0000000000001495

Bressa, C., Bailen-Andrino, M., Perez-Santiago, J., Gonzalez-Soltero, R., Perez, M., Montalvo-Lominchar, M. G., . . . Larrosa, M. (2017). Differences in gut microbiota profile between women with active lifestyle and sedentary women. *PloS One, 12*(2), e0171352. doi:10.1371/journal.pone.0171352

Claesson, M. J., Jeffery, I. B., Conde, S., Power, S. E., O'Connor, E. M., Cusack, S., . . . O'Toole, P. W. (2012). Gut microbiota composition correlates with diet and health in the elderly. *Nature, 488*(7410), 178–84. doi:10.1038/nature11319

Jackson, M. A., Jeffery, I. B., Beaumont, M., Bell, J. T., Clark, A. G., Ley, R. E., . . . Steves, C. J. (2016). Signatures of early frailty in the gut microbiota. *Genome Medicine, 8*(1), 8. doi:10.1186/s13073-016-0262-7

Kelaiditi, E., Jennings, A., Steves, C. J., Skinner, J., Cassidy, A., MacGregor, A. J., & Welch, A. A. (2016). Measurements of skeletal muscle mass and power are positively related to a Mediterranean dietary pattern in women. *Osteoporosis International: a Journal Established as Result of Cooperation Between the European Foundation for Osteoporosis and the National Osteoporosis Foundation of the USA, 27*(11), 3251–60. doi:10.1007/s00198-016-3665-9

O'Sullivan, O., Cronin, O., Clarke, S. F., Murphy, E. F., Molloy, M. G., Shanahan, F., & Cotter, P. D. (2015). Exercise and the microbiota. *Gut Microbes, 6*(2), 131–36. doi: 10.1080/19490976.2015.1011875

Rankin, A., O'Donavon, C., Madigan, S. M., O'Sullivan, O., & Cotter, P. D. (2017). 'Microbes in sport' -The potential role of the gut microbiota in athlete health and performance. *British Journal of Sports Medicine, 51*(9), 698–99. doi:10.1136/bjsports-2016-097227

Steves, C. J., Bird, S., Williams, F. M., & Spector, T. D. (2016). The Microbiome and Musculoskeletal Conditions of Aging: A Review of Evidence for Impact and Potential Therapeutics. *Journal of Bone and Mineral Research: the Official Journal of the American Society for Bone and Mineral Research, 31*(2), 261–69. doi:10.1002/jbmr.2765

Villa, C. R., Ward, W. E., & Comelli, E. M. (2017). Gut microbiota-bone axis. *Critical Reviews in Food Science and Nutrition, 57*(8), 1664–72. doi:10.1080/10408398.2015.1010034

Wang, J., Wang, Y., Gao, W., Wang, B., Zhao, H., Zeng, Y., . . . Hao, D. (2017). Diversity analysis of gut microbiota in osteoporosis and osteopenia patients. *PeerJ, 5*, e3450. doi:10.7717/peerj.3450

Weaver, C. M. (2015). Diet, gut microbiome, and bone health. *Current Osteoporosis Reports, 13*(2), 125–30. doi:10.1007/s11914-015-0257-0

13. TOO CLEAN, OR NOT TOO CLEAN: ENVIRONMENTAL MICROBES

Aliyu, S., Smaldone, A., & Larson, E. (2017). Prevalence of multidrug-resistant gram-negative bacteria among nursing home residents: A systematic review and meta-analysis. *Am J Infect Control, 45*(5), 512–18. doi:10.1016/j.ajic.2017.01.022

Andersen, B., Frisvad, J. C., Sondergaard, I., Rasmussen, I. S., & Larsen, L. S. (2011). Associations between fungal species and water-damaged building materials. *Appl Environ Microbiol, 77*(12), 4180–88. doi:10.1128/AEM.02513-10

Berg, G., Mahnert, A., & Moissl-Eichinger, C. (2014). Beneficial effects of plant-associated microbes on indoor microbiomes and human health? *Front Microbiol, 5*(15). doi:10.3389/fmicb.2014.00015

Bloomfield, S. F., Stanwell-Smith, R., Crevel, R. W. R., & Pickup, J. (2006). Too clean, or not too clean: the Hygiene Hypothesis and home hygiene. *Clinical & Experimental Allergy, 36*(4), 402–25. doi:10.1111/j.13652222.2006.02463.x

Cardinale, M., Kaiser, D., Lueders, T., Schnell, S., & Egert, M. (2017). Microbiome analysis and confocal microscopy of used kitchen sponges reveal massive colonization by Acinetobacter, Moraxella and Chryseobacterium species. *Sci Rep, 7*(1), 5791. doi:10.1038/s41598-017-06055-9

David, L. A., Materna, A. C., Friedman, J., Campos-Baptista, M. I., Blackburn, M. C., Perrotta, A., . . . Alm, E. J. (2014). Host lifestyle affects human microbiota on daily timescales. *Genome Biology, 15*, R89. doi:10.1186/gb-2014-15-7-r89

Hoisington, A. J., Brenner, L. A., Kinney, K. A., Postolache, T. T., & Lowry, C. A. (2015). The microbiome of the built environment and mental health. *Microbiome, 3*(60). doi:10.1186/s40168-015-0127-0

Koljalg, S., Mandar, R., Sober, T., Roop, T., & Mandar, R. (2017). High level bacterial contamination of secondary school students' mobile phones. *Germs, 7*(2), 73–77. doi:10.18683/germs.2017.1111

Lax, S., Smith, D. P., Hampton-Marcell, J., Owens, S. M., Handley, K. M., Scott, N. M., . . . Gilbert, J. A. (2014). Longitudinal analysis of microbial interaction between humans and the indoor environment. *Science, 345*(6200), 1048–52. doi:10.1126/science.1254529

Meadow, J. F., Altrichter, A. E., & Green, J. L. (2014). Mobile phones carry the personal microbiome of their owners. *PeerJ, 2*, e447. doi:10.7717/peerj.447

Peccia, J., & Kwan, S. E. (2016). Buildings, Beneficial Microbes, and Health. *Trends Microbiol, 24*(8), 595–97. doi:10.1016/j.tim.2016.04.007

Riddle, M. S., & Connor, B. A. (2016). The Traveling Microbiome. *Curr Infect Dis Rep, 18*(9), 29. doi:10.1007/s11908-016-0536-7

Rothschild, D., Weissbrod, O., Barkan, E., Kurilshikov, A., Korem, T., Zeevi, D., . . . Segal, E. (2018). Environment dominates over host genetics in shaping human gut microbiota. *Nature, 555*(7695), 210–15. doi:10.1038/nature25973

14. THE FOUNTAIN OF YOUTH *IS* FULL OF MICROBES: THE FUTURE OF THE WHOLE-BODY MICROBIOME

Biagi, E., Candela, M., Turroni, S., Garagnani, P., Franceschi, C., & Brigidi, P. (2013). Ageing and gut microbes: perspectives for health maintenance and longevity. *Pharmacological Research, 69*(1), 11–20. doi:10.1016/j.phrs.2012.10.005

Blaser, M. J. (2014). *Missing Microbes: How the Overuse of Antibiotics is Fueling Our Modern Plagues.* New York, NY: Henry Holt and Company.

Brandt, L. J. (2017). Fecal Microbiota Therapy With a Focus on Clostridium difficile Infection. *Psychosomatic Medicine, 79*(8), 868–73. doi:10.1097/PSY.0000000000000511

Brüssow, H. (2013). Microbiota and healthy ageing: observational and nutritional intervention studies. *Microbial Biotechnology, 6*(4), 326–34. doi:10.1111/1751-7915.12048

Buettner, D. (2008). *The Blue Zones: Lessons for Living Longer From the People Who've Lived the Longest.* Washington, DC: National Geographic Society.

Cockburn, D. W., & Koropatkin, N. M. (2016). Polysaccharide Degradation by the Intestinal Microbiota and Its Influence on Human Health and Disease. *Journal of Molecular Biology, 428*(16), 3230–52. doi:10.1016/j.jmb.2016.06.021

Fond, G., Boukouaci, W., Chevalier, G., Regnault, A., Eberl, G., Hamdani, N., . . . Leboyer, M. (2015). The "psychomicrobiotic": Targeting microbiota in major psychiatric disorders: A systematic review. *Pathologie-Biologie, 63*(1), 35–42. doi:10.1016/j.patbio.2014.10.003

Gibson, G. R., Hutkins, R., Sanders, M. E., Prescott, S. L., Reimer, R. A., Salminen, S. J., . . . Reid, G. (2017). Expert consensus document: The International Scientific Association for Probiotics and Prebiotics (ISAPP) consensus statement on the definition and scope of prebiotics. *Nature Reviews. Gastroenterology & Hepatology, 14*(8), 491–502. doi:10.1038/nrgastro.2017.75

Han, B., Sivaramakrishnan, P., Lin, C. J., Neve, I. A. A., He, J., Tay, L. W. R., . . . Wang, M. C. (2017). Microbial Genetic Composition Tunes Host Longevity. *Cell, 169*(7), 1249–62 e1213. doi:10.1016/j.cell.2017.05.036

Hod, K., & Ringel, Y. (2016). Probiotics in functional bowel disorders. *Best Practice & Research. Clinical Gastroenterology, 30*(1), 89–97. doi:10.1016/j.bpg.2016.01.003

Horvath, A., Leber, B., Schmerboeck, B., Tawdrous, M., Zettel, G., Hartl, A., . . . Stadlbauer, V. (2016). Randomised clinical trial: the effects of a multispecies probiotic vs. placebo on innate immune function, bacterial translocation and gut permeability in patients with cirrhosis. *Alimentary Pharmacology & Therapeutics, 44*(9), 926–35. doi:10.1111/apt.13788

Hungin, A. P. S., Mitchell, C. R., Whorwell, P., Mulligan, C., Cole, O., Agreus, L., . . . de Wit, N. (2018). Systematic review: probiotics in the management of lower gastrointestinal symptoms - an updated evidence-based international consensus. *Alimentary Pharmacology & Therapeutics, 47*(8), 1054–70. doi:10.1111/apt.14539

Konig, J., Siebenhaar, A., Hogenauer, C., Arkkila, P., Nieuwdorp, M., Noren, T., . . . Brummer, R. J. (2017). Consensus report: faecal microbiota transfer - clinical applications and procedures. *Alimentary Pharmacology & Therapeutics, 45*(2), 222–39. doi:10.1111/apt.13868

Lewis, B. B., & Pamer, E. G. (2017). Microbiota-Based Therapies for Clostridium difficile and Antibiotic-Resistant Enteric Infections. *Annual Review of Microbiology, 71*, 157–78. doi:10.1146/annurev-micro-090816-093549

Lurie, I., Yang, Y. X., Haynes, K., Mamtani, R., & Boursi, B. (2015). Antibiotic exposure and the risk for depression, anxiety, or psychosis: a nested case-control study. *The Journal of Clinical Psychiatry, 76*(11), 1522–28. doi:10.4088/JCP.15m09961

McCarville, J. L., Caminero, A., & Verdu, E. F. (2016). Novel perspectives on therapeutic modulation of the gut microbiota. *Therapeutic Advances in Gastroenterology, 9*(4), 580–93. doi:10.1177/1756283X16637819

Pinto-Sanchez, M. I., Hall, G. B., Ghajar, K., Nardelli, A., Bolino, C., Lau, J. T., . . . Bercik, P. (2017). Probiotic Bifidobacterium longum NCC3001 Reduces Depression Scores and Alters Brain Activity: A Pilot Study in Patients With Irritable Bowel Syndrome. *Gastroenterology, 153*(2), 448–59 e448. doi:10.1053/j.gastro.2017.05.003

Rees, T., & Blaser, M. (2016). Waking up from antibiotic sleep. *Perspectives in Public Health, 136*(4), 202–4. doi:10.1177/1757913916643449

Rondanelli, M., Giacosa, A., Faliva, M. A., Perna, S., Allieri, F., & Castellazzi, A. M. (2015). Review on microbiota and effectiveness of probiotics use in older. *World Journal of Clinical Cases, 3*(2), 156–62. doi:10.12998/wjcc.v3.i2.156

Rothschild, D., Weissbrod, O., Barkan, E., Kurilshikov, A., Korem, T., Zeevi, D., . . . Segal, E. (2018). Environment dominates over host genetics in shaping human gut microbiota. *Nature, 555*(7695), 210–15. doi:10.1038/nature25973

Salazar, N., Arboleya, S., Valdes, L., Stanton, C., Ross, P., Ruiz, L., . . . de Los Reyes-Gavilan, C. G. (2014). The human intestinal microbiome at extreme ages of life. Dietary intervention as a way to counteract alterations. *Frontiers in Genetics, 5*(406). doi:10.3389/fgene.2014.00406

Sanders, M. E. (2016). Probiotics and microbiota composition. *BMC Medicine, 14*(1), 82. doi:10.1186/s12916-016-0629-z

Schwiertz, A. (2016). Microbiota of the Human Body: Implications in Health and Disease. *Advances in Experimental Medicine and Biology, 902*. doi: 10.1007/978-3-319-31248-4

Steenbergen, L., Sellaro, R., van Hemert, S., Bosch, J. A., & Colzato, L. S. (2015). A randomized controlled trial to test the effect of multispecies probiotics on cognitive reactivity to sad mood. *Brain, Behavior, and Immunity, 48*, 258–64. doi:10.1016/j.bbi.2015.04.003

Vanegas, S. M., Meydani, M., Barnett, J. B., Goldin, B., Kane, A., Rasmussen, H., . . . Meydani, S. N. (2017). Substituting whole grains for refined grains in a 6-wk randomized trial has a modest effect on gut microbiota and immune and inflammatory markers of healthy adults. *The American Journal of Clinical Nutrition, 105*(3), 635–50. doi:10.3945/ajcn.116.146928

Zhang, C., Li, S., Yang, L., Huang, P., Li, W., Wang, S., . . . Zhao, L. (2013). Structural modulation of gut microbiota in life-long calorie-restricted mice. *Nature Communications, 4*(2163). doi:10.1038/ncomms3163

Index

About the Authors

B. Brett Finlay, PhD, is the Peter Wall Distinguished Professor of Microbiology at the University of British Columbia and a world leader in how bacterial infections work. He has been studying microbes for over thirty years and has published over five hundred scientific articles. A cofounder of the biotech companies Commense, Vedanta, and Microbiome Insights, Brett is Officer of the Order of Canada. He is a co-author of *Let Them Eat Dirt: Saving Our Children from an Oversanitized World*. He lives in Vancouver, BC, with his wife, a pediatrician. They have two adult children, one of whom is his co-author, Jessica.

Jessica M. Finlay, PhD, specializes in environmental gerontology and health geography as a postdoctoral research fellow at the University of Michigan. With degrees from Queen's University and the University of Minnesota, she has won awards and fellowships for her work and authored publications in leading health, geography, and gerontology journals. She is a strong advocate for broadening public awareness of, and developing achievable strategies for, healthier and more inclusive lifelong communities.

wholebodymicrobiome.com